THINKING ABOUT NATURE
An Investigation of Nature, Value and Ecology

Thinking About
NATURE

ANDREW BRENNAN

The University of Georgia Press
Athens

Published in the United States of America in 1988 by
the University of Georgia Press, Athens, Georgia 30602

First published in Great Britain in 1988 by Routledge

ISBN 0-8203-1087-5
ISBN 0-8203-1088-3 (pbk.)

Printed in Great Britain

ERRATUM to *Thinking About Nature*

The publisher regrets the transposition of Figure 4.6 on
p. 61 with Figure 8.1 on p. 126. The captions, however,
are correctly placed.

For Heather, Cally & Nicky

Contents

Contents

Figures and Tables

Figures

Tables

Preface

In this book, I explore the foundations of current thinking on environmental issues. As I show, the ideas on which much of this thinking is based come not only from the sciences but also from philosophy and ethics. Not all writers on the environment and ecology appreciate the many sources of their views, and so I spend some time disentangling the different strands that make up the current consensus position held by many environmentally concerned theorists. The position, although not lacking moral seriousness, turns out to be confused in many respects. A simple, negative aim of the present work is the elimination of such confusion.

On the positive side, I try to contribute to the tradition of thinking about human nature and about the sort of lives that are appropriate for people. Those who live in the world's richest countries are bombarded by glossy images concerning the 'good life'. Advertisements for credit cards seem to suggest that the 'true values' of life include unsecured overdrafts and massive travel accident insurance. Where I live, stores are keeping open longer and longer hours in the evenings, ostensibly for the benefit of customers who are at work during the day. However, it is easy to suspect that they also cater for those who find browsing through shops something to do when there is not much of interest on the television. If it makes sense to think about worth or value in human life, then it makes sense to worry about the emptiness, triviality and banality of life in the consumer society.

The very societies that parade the glossy images before the eyes of consumers are far from well. Their old people are kept sedated in geriatric wards ostensibly because there are no funds to pay for enough nurses to organise suitable, human care and appropriate activities. They display a wide gap between the privileged, and the poor who are exhorted to aspire to levels of wealth that they can never attain; for resources are finite, and there is not enough to keep us all like millionaires. What wealth there is in such countries depends on the existence of a world economic system whose financial power maintains global inequality and drives poorer nations deep into debt. At the same time, money is found for armies, nuclear weapons, the attempted export of military technology to space and the implementation of vast engineering and agricultural projects whose effects on the planet's health are unknown.

If the next century is to see the survival of humans with the prospects of living worthwhile lives, then it is necessary that we ask some relevant questions about our lives and our natures. My suggestion is that in order to discover what sort of human life is valuable we must first consider what kind of thing a human being is. Although there is, in my view, no complete answer to this question, we can — so I argue — grasp one important aspect of human nature by reflecting on what are essentially ecological considerations. The solutions to the problems faced by contemporary society cannot be simply deduced from these reflections. By the end of the book I am able, however, to make some practical suggestions about how to tackle some of our most serious concerns.

As far as possible, the treatment has been kept non-technical. Some aspects of scientific ecology are described in detail in the first part of the book, but mathematical material has been kept to an absolute minimum. For readers who have no background in ecology, I have focused on issues that are thoroughly discussed in modern texts, taking as my standard reference work Begon, Harper and Townsend's recent comprehensive book. Where appropriate, I refer the reader to that work for more information on historical and modern trends in ecology.

A similar approach has been adopted in the case of philosophical technicalities. These have been avoided wherever possible, and my account of moral theory in the later chapters, like the treatment of philosophy of science in the earlier ones, is informal. The result, I hope, is a work that is accessible to readers from a variety of backgrounds, including those who know little of either philosophy or ecology.

Very many colleagues, friends and correspondents have helped with my work on this book. I have learned a great deal from those biologists, ecologists and philosophers whom I consulted and without their contributions the book would have been a poorer one. Among the scientists, Frank Golley and Ian Moffatt were kind enough to let me see unpublished work of which I have subsequently made use. Philip Lobel drew my attention to the latest results on plotting the deep ocean currents and the potential effect of this work on the fishing industry. Among the philosophers, Holmes Rolston and Peter Wenz were generous in providing copies of work, both published and still to be published, to which I refer. I regret, however, that Rolston's new book, *Environmental*

Ethics (Temple University Press, 1988), came to hand too late for the similarities and differences between us on some issues to be discussed in the text. Most of all, I owe thanks to my immediate colleagues at Stirling who have given generously of their advice and time. In particular, Antony Duff drew my attention to relevant work in contemporary ethical theory, while Alan Millar gave a large part of the manuscript a thorough but sympathetic scrutiny and first suggested the title of 'ecological humanism' for the position I adopt. Although the label is not new, it is used by me for a position that differs significantly from the one to which Skolimowski gives the same title (Skolimowski 1981).

For help of a different kind, I am once more indebted to Elspeth Gillespie for her help in typing and revising the manuscript. Jemina Kallas, as copy-editor, ensured a crisp style throughout and Richard Stoneman is owed thanks twice: first, for having commissioned the work (which would certainly not have been written otherwise), and, second, for his patience in the face of my repeated delays in arriving at a final version of the book.

There are some problems about terminology that I have not satisfactorily resolved. When distinguishing the economically richer from the economically poorer countries of the world, I have followed common practice in using terms like 'developed' and 'Third World'. I hope it is clear to readers that I do not regard economically poor societies as necessarily poor in any other respects and that the word 'developed' is not meant to have connotations signifying anything politically or morally desirable.

Introduction

There are so many environmental problems facing us at present that it requires an effort of will to stand back and reflect on them in a philosophical way. It is easy to become so involved in concerns about pollution, loss of species, disruption of biochemical cycles and the thousands of detailed cases falling under these general headings that we do not ask either what *kind* of problems we face, or *why* these problems have arisen. Both of these questions are significant, for in answering them we can start to get clear on some perennial issues of philosophical interest. These issues include the relation of facts and theories, on the one hand, to our values on the other, and also the general questions about human nature and the conditions of human flourishing that have been the concern of nearly all the great thinkers.

One natural way of thinking about many environmental issues is in terms of the technical problems they pose. This can be done through economic analyses of costs and benefits, or engineering challenges, or by reference to sociological and anthropological considerations about traditional ways of life, patterns of settlement and employment. Even this short list allows for numerous technical specialists to have roles in advising those agencies, governments, industries and pressure groups that are likely to be involved where significant environmental impacts are occurring. Strip-mining deep in the Amazon rainforest is a distant, and yet dramatic, example of the sort of project where numerous technical reports can be made, all related to the same project, yet each providing its own perspective. Ecologists who try to assimilate this array of data from different disciplines sometimes despair of finding a technical solution to the issue they confront.

To think of the matter in this way is wrong. Engineering problems have known engineering solutions in many cases, and likewise for some of the others. In certain cases, of course, there is no determinate solution, perhaps because the technical discipline has not developed to a stage at which it can propose them. But suppose that each report commissioned on a certain project both identifies problems and suggests technically feasible solutions. Even if this were to happen, there will inevitably be problems associated with the projects that have not been properly identified or explored in any of the reports. One reason for this

1

is that many of the problems will not be technical ones at all.

If we concentrate on the technical aspects of a project, we are thereby ignoring other ways in which we could be thinking of it. Let me introduce a useful, but purely informal, way of making this point. If we distinguish different *frameworks* of ideas, theories and principles from each other, then we can say that the technical frameworks already mentioned are only a small selection from the number of frameworks available to us when we think about a situation. Even if we take material from more than one framework (combining sociological findings, perhaps, with economic ones) we are still thereby only focusing our attention on certain aspects of the matter before us. I will use the term 'scheme' to signify a composite structure, with elements drawn from more than one framework, using which we can think about a situation.

Frameworks and schemes, as I am using these ideas here, are not meant to be grand things. They are not world-views or *paradigms*. What other authors mean by 'paradigm' is sometimes obscure, although I believe Routley is correct to think of these in general as an attempt to give a unified account of diverse problems and issues as manifestations of one underlying phenomenon (Routley 1983). In much the same way, Kuhn introduced the notion of paradigm somewhat ambiguously into his account of 'normal science' (Kuhn 1970). One idea behind Kuhn's use of the term seemed to be that of a body of achievement, theory and research programmes that would be associated with an enduring group of adherents, and yet remain sufficiently open ended to allow for further articulation. Likewise, a social or cultural paradigm is a shared, underlying perspective on the world and our place in it. A paradigm thus provides a setting for appropriate modes of action, thought and conduct, while these modes are not tightly circumscribed. Within the paradigm, there are opportunities for innovation, freedom of action and development of new themes.

Although different authors tend to come up with different accounts of the contemporary western world-view, the dominant materialistic paradigm or whatever we wish to call it (see the survey in Routley's paper) there seems to me to be some merit in this kind of general thinking. Just as the community in which we grow up sets the standards relative to which we regard beliefs about matters of fact either as obvious or in need of justification, so it very likely sets general value standards of which we are barely aware. Put in these very general terms, current green movements

2

represent attempts to modify the dominant paradigm, replacing it with an alternative which would motivate attitudes, modes of behaviour and dispositions so as to make our society more sustainable, and gentler on the planet.

Although the notions of framework and scheme are not so grand and all-embracing as that of paradigm, they have a use, as I hope to show. Notice, to begin with, that we can diagnose, and provide prescriptions for, problematic situations within the context of some frameworks. We need the framework provided by an economic theory in order to discuss economic problems, just as we need some theory of mechanics if we are to discuss, diagnose and understand problems of mechanics. Not all the frameworks and schemes in terms of which we think are so precisely mapped out as these. Much of our thinking about issues such as education, leisure, the ethics of hunting, the morality of interpersonal relations and so on are done in terms of schemes that are not well-defined. In some of these cases, our very thinking and discussion may be part of a construction process, aimed at developing frameworks within which to think about the subject matter.

Even where a scheme or framework is well-defined, it may not — as a matter of fact — be brought to bear on some issue. Ecologists, for example, may not be asked for an opinion on a project that is of some environmental significance. Since the frameworks and schemes we use in our ponderings and discussions will often guide our decisions, our conduct and our overall attitudes, it is important to be clear on some general features they display. The following four seem worthy of attention:

1. There are an indefinite number of frameworks that can be brought to bear on human life, conduct and society
2. When we restrict our modes of thinking to just one framework, we thereby choose to ignore the perspective supplied by other relevant frameworks
3. Except when dealing with frameworks of the same sort, no straight choices can be made among frameworks. So although one economic framework may be thought superior to another, there can be debate about the relative weightings to be given to physical, economic and political ones
4. The weighting given to one framework for approaching a given subject matter can always be challenged on the grounds that some other relevant scheme or framework has been ignored or weighted too lightly

These may seem somewhat loosely formulated principles, but it should be kept in mind that my approach here is deliberately informal.

Since human nature is a complex subject, it can be discussed and studied through many different schemes. My conception of frameworks and schemes is less grand than the concept of paradigms; but is also wider than Stevenson's characterisation of theories of human nature (Stevenson 1974): for example, although the theory of evolution would count as a framework (in the sense I give), within which to think of human nature, it would not constitute a theory of human nature for Stevenson, for it lacks diagnosis of a 'predicament' and a recipe for getting out of that predicament. By contrast I would count a religious stance like Christianity as a scheme, while Stevenson counts it as a theory of human nature.

Now many of the traditional theories of human nature strike the modern reader as clearly inadequate. Plato's ideal of a state ruled by philosopher-kings involves a very limited account of human potential. For example, the wise rulers in Plato's scheme are meant to prefer a life of philosophic reflection and contemplation, although they recognise the need to attend to the affairs of the state they are governing. The modern reader is immediately struck by the narrow, intellectualised conception of human nature that is required in order to make sense of this preferred form of life. Aristotle likewise can be criticised for maintaining that happiness in the highest sense involves the life of contemplation. But both Plato and Aristotle did recognise to some extent the complexity of human beings and human society. Neither of them thought that all of us could aspire to the ideal forms of human life, but instead they believed that we could live as useful citizens and relatively happy beings when we found a certain kind of balance in our lives. For Aristotle, this involved moderation, including moderation in consumption, and for Plato this involved harmony among the elements of the soul — desire, reason and spirit.

A long tradition of western philosophy has followed the lead of Plato and Aristotle, seeking to do justice to the complexity of human nature. However, these days a different kind of view is in vogue. To get a feeling for it, consider the following line of argument. Let us admit that humans are extremely complicated beings, whose motives, projects, desires, needs and fulfilments are not easily unravelled. Instead of trying to devise an impossibly complex scheme that leaves out the contributions of no relevant

4

framework, why do we not simply look for those forms of social organisation that will permit the fullest range of freedom to persons to pursue their own good and do their own thing? The problem is then to find a means of affording maximum opportunity to all, compatible with protecting certain basic rights and freedoms within the same group. Modern versions of *social contract theory* pursue this apparently liberal ideal, avoiding any attempt at characterising human nature in detail, or prescribing specific routes to happiness and the good life. The proponents of such theories can argue that they are appropriate to democratic, open societies, within which different groups of people can legitimately hold and pursue different conceptions of what is good for humans.

This kind of liberal approach to the problems of human nature, society and morality seems to fit well with the first three principles about frameworks stated above. It recognises the complexity of humans taken as the subject matter of our inquiries; it recognises the bias that is likely to emerge from claiming that things like pleasure, contemplation or consumption of economic goods are the supreme goods in human life, and it notes the incommensurability of alternative schemes. None the less, this approach needs to be challenged, in my view, and challenged for a reason connected with the fourth principle above.

It seems clear that economic, biological or psychological accounts of human nature serve some purpose. I would suggest that they make use of frameworks which supply profiles, two-dimensional portraits, of human nature. As an economic being, some of my behaviour — my economic behaviour — can be described in economic terms. Equally, as a product of natural selection, some of my morphological characteristics and traits can be associated with the history of past adaptation among members of my species to various environments. Neither of these accounts compete with each other, for I am neither a purely Darwinian, nor a purely economic being (principle three above). A scheme based on traditional approaches to theories of human nature might be devised in order to unify the evolutionary, the economic, the political and so on, in one larger story. One such scheme, for example, is sociobiology, about which I have more to say in the first chapter. But if principles one and four above are accepted, then such stories are always doomed to fail. For there will always be the possibility that, however comprehensive the scheme, it can be challenged by citing a framework whose contributions are relevant, and have been ignored. This problem is worsened if we observe that there is also the possibility of different frameworks delivering accounts that compete with, or even contradict, each other.

To avoid these problems, liberal theorists give only the sketchiest accounts of human nature. As will be shown later, such accounts cannot really be taken seriously, but their very sketchiness amounts almost to the adoption of no framework at all within which to consider human nature. But now the very basis of the liberal approach can be criticised on the grounds that it ignores relevant considerations about what kind of beings we are and what sort of life is appropriate to us. My general complaint against the liberal contractarian view is that it is too liberal. This does not mean that I will be urging some elaborate scheme within which human nature can be depicted in all its multi-dimensional glory. Rather, my more modest brand of *eco-humanism* claims that, among the relevant frameworks that political and ethical thinking ought to use is one deriving from scientific ecology.

My brand of eco-humanism contrasts strongly with various trends in contemporary environment movements and environmental philosophy. One particularly interesting position, or platform, is the one associated with the *deep ecology* movement. Like other environmental movements, the deep ecologists argue that human flourishing or self-realisation requires a re-evaluation of our relationship with the rest of nature. This re-evaluation has taken various forms in the brief history of the movement, but a prevailing theme has been to urge the abandonment of our human-centred modes of thinking and valuing, and — more recently — to undertake a real identification with nature. These ideas form a major topic of investigation in the second part of the book.

There is, however, a complication which affects not only the deep ecology position, but also the kind of position taken by other contemporary writers on environmental ethics. Although no matter of fact can justify a moral judgement, there is some connection between the factual and the evaluative. Any moral code draws on the context within which it is held, and this context will include not only conceptions of the right way for humans to develop, and notions about the appropriate goals for human society, but also a conception of what humans are and what kind of world they inhabit. This latter conception is liable to be influenced by a background stock of 'common sense' beliefs prevalent within a society as well as by religious, philosophical and scientific ideas that are current.

Many contemporary ethical positions draw support from certain philosophical and scientific models. This is true of deep ecology, whose supporters appeal often to certain metaphysical frameworks. Two common appeals are to idealism (the claim that the world is in some way mind-dependent) and to various kinds of global holism (the idea that all things are interdependent in a significant way). These metaphysical positions are sometimes associated with contemporary physics — as, for example, in the work of Fritjof Capra. I will be maintaining that physics gives no special support either to idealism or global holism and that these doctrines should be recognised for what they are — metaphysical positions that are not open to conclusive proof or refutation.

An additional background stance that is also associated with deep environmental positions concerns the nature of *explanation*. It is sometimes argued that scientific, ecological studies show that traditional atomistic, and reductive modes of explanation have to be replaced by non-reductive, holistic modes. I will be arguing in some detail that scientific work in ecology gives no real support to such claims. Much of the focus of contemporary ecology is adequately described as seeking the same kinds of explanation as are sought in other sciences. This particular grounding for the new approach to value theory is thus flawed.

If deep ecology and related ethical positions lack a grounding or rationale, we might wonder if their value recommendations can be saved. In fact, there are ways, as I will argue, of reconciling my variety of eco-humanism with much — though by no means all — of the value-content of these views. The project of the book is thus brought to a conclusion by advocating a means of limiting the liberal, contractarian account of morality by taking account of what scientific ecology does reveal to us about human beings. What is revealed is not an obscure metaphysics, nor a new way with explanation. Instead, it requires us to take seriously the notion that we are a part of nature, and that — to put it crudely — what we are and ought to be is partly determined by where we are.

For convenience, the book is divided into two main parts. After the initial two chapters that develop the idea of thinking in frameworks, the reader is introduced to ecology and its history. This permits an informed discussion of the issues of reduction, holism and explanation. As well as negative points of the kind already mentioned, I try to make clear just what the truth is in the claim that what we are depends on where we are. Tackling this claim involves some realignment of traditional notions about our nature, and about what is essential to that nature.

The second part of the book, from Chapter 9 on, considers the nature of ethical theory, the strengths and weaknesses of human-centred ethical perspectives and the prospects for eco-humanism. The impact of eco-humanism on our normal modes of political thinking is studied, and a brief attempt made to clarify some tricky issues (like vegetarianism). If eco-humanism is taken seriously, we need also to consider the kind of education appropriate to our situation, and to the situation of those who live in the most threatened parts of our planet. My suggestions towards the end of the book are out of step with much of the ideology of the contemporary western world. I hope this fact alone does not diminish their interest.

1

Thinking about the Environment

1.1 Reality and ideals

We live in a world that supports a large human population, currently in the region of five billion. Recent estimates suggest that by the year 2025, this will have risen to around 8.2 billion, the largest growth coming from urban populations in the Third World. Since, as far as we know, human life is distinguished from the rest of animal and mammalian life by manifesting self-reflective, high intelligence, and since many of the people now living can expect to be alive in 2025, it is interesting to wonder whether humans living then will be offered a higher quality of life than is offered just now.

Engaging in this sort of speculation leads to two extreme results — optimistic or pessimistic. Optimistic, utopian thinkers look forward to a world where more life is linked with better life. Thanks to the benefits of cheaper energy, the fruits of biotechnology and selective plant breeding, more enlightened international political co-operation and widescale education, the humans living in 2025 can, on this scenario, enjoy an unrivalled quality of life.

What is meant here by 'quality of life'? I think that it means at least a life in which basic material benefits and freedoms are well established. Among the benefits are adequate food, education, health care, a basic standard of housing, provision of adequate fuel for cooking, access to activities that are of a life-enhancing sort — sport, entertainment, forms of cultural and community activity and so forth. Among the freedoms are freedom from hunger, ignorance, illiteracy, the fear of treatable diseases, and from the worst indignities of poverty; and freedom

9

to make a certain number of personal choices, ranging from what to wear or eat, to what — if any — religion to practise. To those currently living in the rich, or developed countries, such a list of benefits and freedoms may seem trivial. But a world in which such benefits and freedoms were widely dispersed (they are unlikely to be universal) would be a significant improvement on the one we currently inhabit.

Notice that in dealing with quality of life in this way, I have given little emphasis to many of the so-called 'higher' things in life. I have not mentioned high culture, the life of the mind, or spiritual values other than those associated with religion. Later, we will consider the issue of quality of life in more detail, concentrating on the question of what makes a human life valuable, or worth living. For the time being, it is just as well to focus on what those living in developed countries conceive as being the ordinary, or the banal. For the very ordinariness of such things can easily lead us to overlook them. Let us be realistic enough to recognise that the widespread availability of such basic goods would be a considerable achievement.

A world in which the population of humans may have increased by 60 per cent and where these basic benefits and freedoms have become widely dispersed might be thought to be a better world than this. For, at least on *utilitarian* grounds, the more benefits there are around, the better. There are many objections to such a line of reasoning. But instead of thinking about them at the moment, let us consider a more urgent question. Assuming the truth of the utilitarian thesis, have we any reason to be optimistic about the future in the way imagined?

One reason for pessimism comes from reflecting on the past history of human achievement. Think of our technical innovations, for a moment. There seems little doubt that the basic quality of life of some people has been improved by the advances in technology in the industrialised countries. The liberation brought to ordinary people by the invention of the internal combustion engine and the subsequent development of the motor car is an obvious example. Mobility and freedoms beyond the dreams of our predecessors have come to people in those countries where car ownership is widespread.

Such technological progress, however, has been purchased at a cost. Motor cars are major polluters in Europe, and share a considerable responsibility for acid rain, and the elevated levels of lead in the blood of children. They also lead to the deaths and maiming of thousands each year. They now congest what used to be pleasantly quiet village streets and town squares. The motorist's freedom to travel at will has also reduced many other freedoms, for example the freedom of children to play safely in the side streets, and a considerable amount of time must now be invested in teaching the rudiments of road safety.

The story just told about cars can be repeated for other technological innovations, even ones as simple as the canning of food. Benefits have always incurred costs, and an interesting question can always be asked about the relative weighting of costs and benefits. As we will see later in this chapter, such weighting is no easy matter, for there is no objective standard we can use. To judge by their behaviour, those people who live in societies in which car ownership is widespread are usually pretty certain that the benefits of the motor car outweigh its costs, for there is no shortage of purchasers and — with only a few exceptions — most people in such societies would like to own a car. So it is with the convenience of electricity, the benefits of which are commonly perceived as outweighing the short- and long-term costs associated with its production — for example, air and estuarine pollution from coal- and oil-fired power stations, radiation risks and spent fuel disposal problems in connection with nuclear stations.

Such reflection on costs and benefits, however, does little to determine whether our outlook for the future should be pessimistic or not. On the whole, we might say, things have not worked out so badly. For each cost that the pessimist fastens on, there are benefits that the optimist can indicate. No-one knows for sure how intractable the problem of nuclear waste disposal will prove to be; but the benefits of nuclear power are enjoyed by millions of those who use electricity every day for cooking, heating, lighting, entertainment, not to mention the massive industrial users of high-quality energy. No-one knows for sure that soil degradation, erosion and loss of productivity are the inevitable outcomes of high-input farming methods, widespread adoption of monoculture and commitment to high-yield crop varieties. Yet again the benefits of modern agriculture are not only undeniable, but taken for granted in the industrialised countries.

Where then, is realism to be found? If we try to steer a course between the twin perils of utopian optimism and doomsday prophecy, what assessment of the global situation do we endorse? In particular, are we faced with an environmental crisis? Are drastic steps urgently required to save the planetary ecosystems from imminent collapse? Are the benefits of technology, modern farming and industrialisation clearly outweighed by the costs? My answers to the last three questions are Yes, No and Perhaps respectively. Yes, we are indeed faced with an environmental crisis — but one for which modern technology is not solely answerable, for its origins go back hundreds of years. Indeed, despite the pollution of drinking water associated with the electronics industry in California's Silicon Valley, many modern industries are environmentally kinder than their heavy industrial ancestors. Despite the disappointing performance of the 'Green Revolution', modern agriculture is probably no more damaging to the soil than much of the husbandry practised in ancient times. The problem here, as elsewhere is to a large extent one of *scale*. Although there are environmental problems that need to be tackled, these have been building up for some time and with luck have not yet reached catastrophe point. The very scale of our environmental problems means that it will be a sensible strategy to try and tackle them after considered thought instead of letting ourselves be panicked into precipitate action.

None of what I am saying here is meant as a recipe for inaction, or as bland optimism. I am not in the business of commending the 'technological fix', that is, the idea that to each problem posed by the advance of technology there will be a solution found by further applications of technology. But, at the same time, I am not in the business of recommending that we abandon the benefits that industrialisation and technology have brought us in order to pursue some ecotopian ideal. My business in this book is to consider our attitude to nature, the moral standing, if any, of other natural things apart from ourselves, and the question of whether the biological sciences — in particular ecology — can give us the insight or information that may help us plot a sensible strategy for our future dealings with nature. That in itself is a big enough project without taking on the further challenge of defining and recommending brand new lifestyles for the twenty first century.

1.2 Thinking in frameworks

I have already claimed that we have an environmental crisis on our hands. In my view, this crisis has not come about because people these days are doing anything very different from what they have been doing for the last two or three thousand years. If any one factor is to be blamed, it is technology, closely followed by industrialisation and the economic systems of major national states. Thanks to the technologies we now possess, our activities have reached a scale where they are having a global impact of terrifying proportions. What may seem to be our 'natural' human modes of behaviour are now having extremely rapid and widespread effects on chemical cycles, the lives of other species and the whole tissue of planetary life-support mechanisms. However, we still have time to do something about the impact of our activities — if we so wish. There is nothing inevitable about environmental degradation. But, as far as we know, the present age is the first one in which human beings have been faced with the choice of whether to live and let live, or whether to continue with patterns of behaviour that will lead to the elimination of many other forms of life.

It is important to recognise, however, that the very idea of there being an environmental 'crisis' involves more than the detailing of facts. We can point to the build up of various chemical agents in food webs, indicate the loss of six million hectares of land each year, note the felling or burning of around eleven million hectares of forest each year, calculate the cost of acid rain damage to buildings and forests, and so on (WCED 1987; Holdgate, Kassas and White 1982). But no amount of dwelling on these facts will establish that we are faced with a crisis. For there to be a crisis something has to be wrong — it has to be the case that these things should not be happening. And to explain why these things are wrong, is not such an easy matter.

One problem is that things are not right or wrong in isolation. Sometimes, as in the case of table manners, or the side of the road on which to drive, it is simply a matter of social convention that determines what is right and wrong. So there is nothing intrinsically wrong about driving on one or another side of the road — this is a matter that inherits rightness or wrongness from a context. Nor, although most social contexts agree on this, is there anything intrinsically right about honouring parents, or giving respect to those older than oneself. Even if we failed to find an actual society in which parents are not honoured by their children, we can at least engage in fantasy. We could imagine a society within which there are moral norms, and social conventions, but in which honour for parents was neither. Such a society would not be immoral, simply different from any that we happen to know about.

13

Consider the question of whether it is right or wrong to educate children in a certain way. Suppose, for example, we think of 'traditional' forms of American or European education. In such forms, the children are given minimum freedom in school, they are relatively passive recipients of information, and they are very much under the control of the teacher. To caricature, the teacher is active, powerful, the person who gives knowledge to an inferior. Suppose we go along with critics of traditional education. We accept, then, that such education does not turn out adults who are active, inquiring, ready to participate in communal activities, self-reliant, willing to criticise themselves or others. Instead, it by and large turns out adults who are passive, accepting of external authority, and — in combination with a certain kind of assessment procedure — it teaches the bulk of those who pass through the system that they are failures, persons of little intellectual or cultural worth. By contrast, a small number of children do rather well under this same system, and these are the ones privileged to receive positions of power or influence in the society which they enter as adults.

Accepting all this, nothing so far follows about the rightness or wrongness of such an educational system. Notice that I am not discussing the question whether traditional education really does deserve this characterisation. Rather, my point is more radical. Even if we accept the description, we do not have, so far, any reason for setting out to reform such an education system. For — as Dewey very clearly saw at the turn of the present century — to see such a system of education as wrong is something that requires appeal to some larger political or moral considerations about the society within which the system is practised. For Dewey, the traditional education was wrong — among other things — because it failed to live up to the democratic ideal. But the democratic ideal is one that is appropriate to a certain kind of society (Dewey 1916: Chs 1–4 and pp. 321–4).

What Dewey in effect did was criticise traditional education because of its failure to produce citizens who would contribute conscientiously to a democratic society. Speaking loosely, we could say that he saw a 'contradiction' between the ideals of American society and the kind of education taking place in traditional schools. He diagnosed an educational crisis, because he had a perspective on society according to which schools should be producing self-respecting, critical, active, participating citizens. An education system that undermined the pupils' self-respect, which encouraged passivity and deference to arbitrary authority (rather than to the authority of reason, and the norms of inquiry) was wrong for a democratic country (Dewey 1915, 1916). This diagnosis on his part shows clearly that he was locating a problem within a larger framework; without this location, or some other framework, to guide us we would not be able to identify traditional education as problematic at all.

Of course, someone might point out that it is wrong in any case to treat children merely as passive recipients of an education process. But if we ask why this should be, we come back again to a larger framework of ideas, a perspective within which this judgement fits. It may be that we regard all persons, regardless of age, and regardless of social context as having the right to certain sorts of treatment. Such treatment would avoid manipulating them, or encouraging undue respect for arbitrary authority, and so on. This time, the objections to the education system arise not from within a political framework but from the perspective of a moral framework, one that operates with a certain conception of human beings, their natures and their rights.

We can observe, then, that the general frameworks of ideas within which we can think of, diagnose and propose solutions for problems need not be of one kind. More than one framework can be called upon when we think about an issue. Sometimes, as in the case just described, two frameworks will deliver the same verdict — that a certain procedure, or institution is wrong. This is a case of overdetermination — where the same verdict is given from more than one perspective. In other cases, we may expect different frameworks to deliver inconsistent verdicts. Thus economic and religious frameworks may come up with very divergent evaluations of the same course of action. In such cases, we face interesting choices — choices as to what perspective we are going to adopt, and what grounds there are for preferring one perspective to another.

The frameworks used in philosophy have a feature worth noting. Just as humans reflect on their own situation, so in philosophy we bring frameworks to bear on other frameworks. There is nothing very puzzling about this, just as it is not surprising that in natural languages we are able to write about natural language. (Such a practice may be a source of paradox, but that is a separate issue.) Some of what I do in this book thus constitutes a study of first-order frameworks by means of what might be called *second-order* frameworks.

15

1.3 Sociobiology

Those who think the technological fix will solve all of our problems might be expected to think that the sciences provide the best frameworks within which to think about world affairs, the human condition and the prospects for sustainable development. Yet, the attempt to view the world in the detached, objective terms of the sciences is not simply the prerogative of technological optimists. Indeed, we face a paradox when we think about humans and the natural sciences. On the one hand, human beings seem very much a part of nature, their bodies simply physical systems among other physical systems. On the other hand, we seem distinct from, even transcending, nature, for each of us is a unique centre of self-reflective consciousness, a being with a distinctive point of view, a subject with feelings and even with feelings about its feelings. Are we in some way in, but not of, nature? Or are we simply ignorant of the right naturalistic descriptions to use in giving accounts of human intention, motivation and action?

These problems about the scope of the scientific enterprise and its ability to deal with the human subject (who is also an object for part of the enterprise) are quite distinct from other aspects of the sciences that are of relevance to environmental thinking. The physical and biological sciences are rich in suggestions for ways in which we might tackle various problems. With suitable restraint, appropriate or intermediate technology has the potential to solve many problems. We know how to provide cheap accommodation, wholesome food, clean water, basic sanitation and simple cooking facilities for those who are presently lacking all, or some, of these things (George 1977, 1984; Tudge 1977). What we seem to lack is the political or practical drive to tackle the problem of providing these. A relatively minor reduction in the defence budgets of some of the larger world economies would easily finance a massive improvement in the lives of the world's underprivileged. Were we to act in this way, the physical and biological sciences and their associated technologies would be ultimately the source of our ideas, our techniques and our practical methods for tackling these projects.

The impacts of the sciences on our ways of thinking and acting are thus likely to be various. But let us stay, for a moment with the questions raised by the attempt to think about humans and their problems within scientific frameworks. First of all, let us think about the issue of human freedom, or *autonomy* (to use the grandiose term of the moral philosopher). On one view, humans and other animals are radically distinct. Non-human animals, some theorists tell us, are objects, not subjects, responding to environmental stimuli, rather than initiators of deliberate, intentional action. They live in nature, while we have culture (Freire 1974: 3–4). There seems to be some truth in this. Even though human actions are by and large stable, and predictable, it hardly strikes us when we think about ourselves in relation to other creatures that we have very much by way of continuity with them. Music, literature, art, the possession of complex linguistic skills, sympathy, the enjoyment of a good joke, the desolation of grief — these are all features of human life and culture that seem to have no real parallel in the lives of non-human beings.

Other theorists, inspired by the last century of biology, have tried to suggest that there are impressive continuities between humans and other living things. Our current knowledge of genetics, coupled with the Darwinian theory of natural selection, suggests that we — and all other living plants and animals — are simply, in Dawkins' phrase, 'survival machines' for genes (Dawkins 1976). Such theorists find in our genes the secrets of our innate wiring, those dispositions with which we are born, which have evolved through the long history of our species, and which exert a powerful influence on our attitudes, desires, social structures, and even on our morality (Wilson 1975). Of course, none of these writers is claiming that we are directly under the control of our genes. Rather, our genes provide us with systems which, programmed with so-called 'epigenetic rules', respond to the environments in which we live, develop and die by inclining us towards certain kinds of behaviours — behaviours which we, from the inside, think about as being under our own control, and of our own creation (Ruse 1986: 143–7). But if the sociobiologist is right, these behaviours are less under our own control, and less subject to rational constraint, than we might think.

Even the toughest of sociobiologists seem to be somewhat repulsed by their own theorising. For the theme of how we are all on genetic leashes is often rounded off by a coda which appeals to a future in which we transcend these narrow, selfish constraints, and strive to build a more altruistic, caring, liberal world than the one coded for in the gene pool. Having told us how we are the products of 'selfish' genes — and having concluded that we are thus ourselves selfish (a staggering *non sequitur*) — Richard Dawkins concludes his popular work on this subject by writing:

A simple replicator, where gene or meme cannot be expected
to forgo short-term selfish advantage even if it would really
pay it, in the long term, to do so . . . our conscious foresight
— our capacity to simulate the future in imagination —
could save us from the worst selfish excesses of the blind
replicators. . . . We can see the long-term benefit of
participating in a 'conspiracy of doves', and we can sit down
together to discuss ways of making the conspiracy work. We
have the power to defy the selfish genes of our birth and, if
necessary, the selfish memes of our indoctrination.
(Dawkins 1976: 215)

This admission makes the whole status of the 'selfish gene' theory
somewhat suspect. For if we are free, all along, to defy the genes
that supposedly control us, then they do not really control us; in
which case, we are different from creatures whose behaviour is
coded by their genes, and whose actions are predictable in a
precise way by those who have unravelled the epigenetic rules for
their behaviour.

I am not out to deal lethal blows to sociobiology. It is clear that
there must be a great deal yet to be learned about human nature
and the biological sciences have surely a large role to play in this.
Sociobiology is perhaps best thought of as a crude first attempt at
articulating the kind of information about human nature that can
be derived within a biological framework. We can accept this even
while recognising that it does so in terms that often seem to reflect
the predispositions of its supporters rather than some neutrally
characterised subject matter. The very vocabulary of 'selfish' and
'altruistic' applied to items like genes (which, since they cannot
think or act, cannot be either selfish or altruistic) reveals this kind
of bias. Further, since the sociobiologist emphasises continuities
between humans and other animals, it is not clear that the
explanatory direction should take us 'downwards'. Instead of
thinking of human behaviour in terms of animal rituals, the
territoriality of the pack, and the dominance hierarchies in the
herd, we might equally well look for human-like features in other
kinds of creature. Assimilation is a two-way business.

These points aside, sociobiology indicates a possible scheme of
thought about *human nature*, although so far not a finished account
of this nature. Now it is important for some of my later ethical
analyses that I make clear just where I think human beings do
stand in nature. Like the sociobiologist, I believe that human

beings are products of just the same kind of selective forces that have produced the other species with which we share the planet. As will be seen in the next chapter, evolutionary explanations fit in nicely with ecological ones to give us a picture of how adaptation to environments works. But the very evolution of human beings has led to the production of creatures of great intelligence, of self-awareness and for whom moral values are significant. I do not claim to know whether the appearance of these specifically human features was accidental, a by-product of the evolution of more reproductively-relevant characteristics or something else. For my purposes, it does not matter that we have no answer to this question. All that I need from the reader is agreement that we are natural beings, and that many of our basic physiological properties and traits are the product of natural selection. As will be shown later, this footing in the natural world is enough to motivate certain kinds of ethical concern about the same world.

It is worth noting, in order to avoid confusion either now or later, that genetic and evolutionary accounts of our nature — at whatever level — do not commit us to the bugbear of *hard determinism* (the doctrine that we have no real freedom of action). No scientific account of human or animal behaviour need frighten us on this account. It is true that at certain very low levels of description — those concerned with fairly basic physiological processes, for example — there is a high degree of determinacy. Without this, bioengineering, gene splicing, vaccination and countless other kinds of procedure would be as likely to fail as succeed. At a different, 'higher' level of description, we are aware that dogs conditioned by certain kinds of training are likely to obey simple instructions, or that alcoholics exposed to a certain regime stand some chance of cure. But our confidence that causal processes are operating at these higher levels is perfectly compatible with our normal beliefs about human and animal agency, in particular, with our beliefs about freedom.

Freedom, as we understand this notion in application to people, and also (though more controversially) in application to other animals like dogs, is the ability to act free from various kinds of compulsion or constraint. But this does not mean that it is freedom from all causal forces. In other words, freedom does not mean randomness, whimsicality, or arbitrariness in reflection, decision or action. Of course, there may be times when we do act in arbitrary, or random manners — when we wonder later why we chose to turn right rather than left, or why we stopped at one point on the walk rather than another. But most of the time, we can make sense of the thought that there are perfectly good causal explanations for our actions, while yet those actions are free.

The position I am taking here is known as *compatibilism* or 'soft determinism' and, as championed by Hume, its defence involves showing that the dispute between the determinist and the libertarian is largely verbal rather than real (Hume 1758: §viii; and 1739: II, iii; a recent defence of the position can be found in Dennett 1984.) The soft determinist does not try to distinguish humans from the rest of nature in terms of a special freedom we enjoy that is denied to other creatures. I am not free from the laws of nature, and I am not able to act totally unfettered by causation. Rather, if I differ from a dog it is not because he is a causally bound creature and I am a free one. Rather, he is a creature of limited intelligence, foresight, deliberation and self-regard. My own powers in these respects are formidable compared to his, but we are both creatures whose behaviour is caused, and whose actions are explicable by reference to the appropriate causal factors.

Any appeal to continuities with the rest of nature will not then excuse damaging behaviour towards the environment. We can hardly expect a beaver to consider the environmental impact of the floods its construction work might induce. Like beavers, we too are creatures whose behaviour is caused. Moreover, we can appeal to biology, psychology and the social sciences to explain why we too wish to engage in construction projects that will lead to the flooding of valleys. But we, unlike the beaver, can take such studies into account — along with everything else — when we give thought to the environmental impact of our actions. Unlike the beaver, we can take responsibility for all the foreseeable consequences of our actions and even chide ourselves for not foreseeing in time some untoward consequence of our project.

2

Frameworks

2.1 Economic, political and moral thinking

It should by now be clear that, in the sense in which I am using the term 'framework', no one framework of ideas is likely to be adequate for the formulation, and solution, of the environmental problems we currently face. Usually a scheme, drawn from a number of frameworks, is brought to bear on any serious issue. Schemes, like frameworks, are seldom adequate for the purposes to which they are put.

A tendency of which we have to be wary is the attempt to adopt just one framework or scheme, and then try to claim that in terms of it we have all the materials for understanding and solving our problems. Unfortunately, a normal tendency, even among sophisticated thinkers, is to do precisely this. Some philosophers, for example, have allied themselves with the sociobiological programme just discussed. For them, the combination of genetics and evolutionary theory provides sufficient richness to characterise the human condition and explain all we need to know about our desires, needs, interests and potential. With great candour, Michael Ruse writes in the following way about his own situation at the time he was asked to give evidence at the Arkansas Creation Trial:

> I grew to realize that at least my Creationist opponents had a sincerely articulated world picture. I had nothing. Even though I had been a professional philosopher for 20 years, I still had no settled thoughts on the foundations of knowledge or morality . . . I have now come to see that our biological origins do make a difference, and that they can and should

be a starting point for philosophy. (Ruse 1986: xii–xiii)

Other thinkers have in a similar way allied themselves with religious or political schemes for which they claim equal scope.

In our cooler moments, away from the enthusiasm of the study or the library, we are aware of the oddness in claims like 'humans are purely political animals', or 'we are nothing but survival machines for our genes'. We realise that humans are biological beings, and also political beings; that they are economic beings and also moral beings; that they are physical beings but also beings who care about the life of the mind and 'spirit'. To some extent, the strengths of economic, political or biological perspectives are best probed and shown by attempts to reduce a whole host of behaviours to description in terms of the favoured framework or scheme. With enough ingenuity, we can start to make a case for the comprehensiveness of any one framework. But this is in the end just an academic exercise. None of us ought to take seriously the claim that *all* that matters in human life can be adequately characterised in purely economic, religious, biological or political terms, or in terms of any other scheme. If I have been somewhat cavalier with sociobiology in the preceding section it is simply because I cannot believe that defenders of sociobiology are doing more than giving the perspective a decent run for its money. Extreme claims about the 'reduction' of one subject matter to another are, in any case, usually based on confusions (as will be shown in Chapter 5).

To make this point clear, consider the application of economics, especially in the form of cost-benefit analyses, to various environmental problems. In order to run such analyses and produce environmental impact assessments, it is necessary to 'reduce' each valued thing to a common measure — monetary value. The valued things can include individual natural objects (human beings, trees, mountains, birds, lakes and valleys), systems and communities containing such objects, buildings and other artefacts, and activities (business, leisure and others). It is well known that so-called 'intangibles' or 'externalities' are particularly hard to value. Such things include human life and health, availability of certain aesthetic experiences, culturally or historically interesting buildings, fresh air, clean water along with most of the other things which are of importance to worthwhile human living.

However, it is also true that for the purposes of a particular analysis we can make carefully weighted assignments of monetary value to such things. In so assigning values we inevitably weight the outcome of the analysis according to the values initially chosen. So the preservation of an industrial landmark (for example, a historic factory building) may be rejected as not cost-effective under one analysis, but accepted under another. The difference resides in the values built in by the varying analysts. Such a result is not surprising. Nor is it a reason for giving up economic valuations. Cost-benefit analyses are perfectly useful tools when their limitations are properly understood. Maybe the amount of time, energy, care and love I invest in my family is not cost-effective according to one analysis. But this shows that such an analysis is inappropriate to family life. When we think of such analyses in connection with things that matter deeply to us, we recognise immediately the inappropriateness of trying to live our lives as purely economic beings. Yet we are, for all that, economic beings, and would expect to be treated as such in the right context. For example, as an employee of a corporation, let us suppose, I would expect the economic costs of employing me to be balanced by measurable benefits to the organisation accruing from my services.

Thus there is really nothing of substance at issue in the question of whether we are economic beings. The unsatisfying, but true, answer is that, from the standpoint of those theories of society that are economic, we are such beings. But no such theory can claim to give more than a partial answer to the question of what human life and human society is all about. A significant, but false, claim is that all aspects of human life can be adequately explained and described in purely economic terms. Likewise, from the standpoint of those theories that are recognised as belonging to biology (evolution, ecology, physiology, and so on) we are biological beings (we are beings who have evolved, who interact with their environments and who can be described in physiologists' terms). The only problem again comes when theorists try to claim comprehensiveness for their own specialist frameworks.

Like the economist, the evolutionary biologist can give a description of what seems to be a pretty non-evolutionary matter. Just as my love for my children can be represented in monetary terms for the purposes of an economic analysis, so a biologist can characterise an act of impetuous generosity as the product of an essentially selfish system, driven by codings in the subject's DNA. Neither the economist nor the biologist are maintaining anything that is unintelligible. It might have been that humans were much less complex creatures than I am taking them to be. The creatures described in economics textbooks or in treatises in sociobiology are pale shadows of real human beings simply because the very constraints of the perspective cut off from our view the emotional, moral and cultural diversity and depth of human nature. This is no objection to these frameworks. Their power is shown by the ability they have to give us a one-dimensional view of the human subject. What we must guard against is the attempt to make the complex and multi-dimensional appear simple and one-dimensional.

Some of those, like the Creationists, who oppose Darwinism in biology, might be concerned not so much by the thinness of its account of our nature but by the thought that it is false. This is a problem that is going to arise for any serious scientific or quasi-scientific attempt to describe the world. It is useful to distinguish the truth or falsity of particular biological, sociological and economic theories from the more general perspective that such sciences provide for our view of the world. Thus, from the point of view of Marxist political-cum-economic theories of society, humans living in capitalist societies are exploited, alienated and oppressed (McLellan 1977, 1980, 1981; Suchting 1983). Now, if Marxism is false, oppression and alienation either do not exist in society or exist in different forms and for different reasons from those identified by the Marxist. My own position here is to regard Marxism as flawed in its details but along the right lines in its approach to the description of society. Marx himself gestured towards what we might call an ecology of society while failing, in my view, to establish that the mode of production in capitalist society is the source of our social evils. To reject Marxism itself is not to reject its theoretical approach, an approach that treats human society as developing under the impact of technology and modes of production. In other words, even if we do not accept the details of Marxist analysis, it is an example of an intelligible perspective on one aspect of human social life, seeing it in relation to the apparatus of ownership, production and labour within society. We have to classify schemes and frameworks into kinds or sorts. One socio-economic framework may be compared with another, even though neither can be readily measured against a framework of a different kind (for example, a physical one).

It is common for those who reject the historical materialism and economic determination of the Marxist view of society to maintain a certain sympathy with its moral critique of capitalism (for example, Stevenson 1974: Ch. 5). And we might think that moral frameworks — those within which humans are seen as creatures motivated by specifically moral concerns — would provide more comprehensive accounts of humanity and society than any of the others we have so far mentioned. Of course, as far as our moral nature is concerned, a moral framework will be pretty comprehensive. But I would argue that such a framework has the same limitations as any other. Humans are not purely moral creatures, and moral considerations are not the only sort that loom large in our thinking. In fact, it is hard to establish that moral considerations loom large in our thinking even on practical issues with a moral dimension. Although moral considerations have a particularly serious force, their force is hardly compelling; nor is it clear that rational beings will always regard the moral considerations in any situation as the overridingly important ones (Williams 1985). In their second-order application, however, moral frameworks are important because they enable us to argue that certain first-order frameworks *ought* to be applied to a problem confronting us. That an ecological or political framework has not been used in thinking about a certain project may be a legitimate moral criticism.

Moral philosophy these days, like other areas of philosophy, is under pressure from the *sceptic*. We live in sceptical times, in which people in the industrial countries of the world have lost faith in many of the values that were traditional in their societies. Those who try to maintain, or defend those values are attacked in basically sceptical ways (I count cynicism as a form of scepticism). The strength of scepticism is that it will take nothing for granted; it asks for justifications for beliefs, however widespread, however basic and however obvious. And it is not possible to supply justifications that measure up to the standards the sceptic demands. So powerful is scepticism these days that several philosophers have suggested that it is simply something with which we have to learn to live (for example, Nagel 1986: Ch. 5). But moral scepticism is something that threatens our ability to go on living in decent, or even sustainable ways. Like the death of God, scepticism in ethics is a condition consequent on which it seems that everything is permitted.

Later we will look in detail at one liberal conception of morality — the conception associated with social contract theory. By this account of morality, we have to think about the basic rules for social living in terms of a fiction. The fiction is the supposition that at some imaginary time people chose freely to group themselves into a society. But since — at the time of choosing — no-one knows just what his or her position in the future society is to be, the persons in the original position have to agree on principles of justice that will guarantee fairness in society to those in all the stations that will be occupied.

It might appear difficult to extract even general principles of justice from such slender beginnings, but Rawls, in following up the implications of bargaining in the original position, arrives at the following two principles:

> First, each person is to have an equal right to the most extensive basic liberty compatible with a similar liberty for others. Second: social and economic inequalities are to be arranged so that they are both (a) reasonably expected to be to everyone's advantage, and (b) attached to positions and offices open to all. (Rawls 1972: Ch. 11)

Here, then is a promising beginning to the moral enterprise. Without specifying any particular style of life, modes of conduct or special duties of humans in society, we have begun to characterise what a fair, decent society would be like. Such a characterisation is particularly suitable for societies that are open and democratic. For such societies permit their members a wide range of options concerning styles of life, religious belief, types of work and forms of leisure which can reflect different valuations of the projects and enthusiasms humans can devise for themselves.

One problem that we will have to face later concerns whether the liberal, open society is the best choice for humans, given the kind of creature they are. A second, related question concerns the status of things other than human beings, given the kind of conception urged by the contract theorist. The *land ethic* put forward by Aldo Leopold in the middle of the present century suggests that not only is there value in things natural, wild and free, but also that we have a relation with the land which is akin to the contractual relationship that binds society. 'That land is a community', he wrote in 1948, 'is the basic concept of ecology, but that land is to be loved and respected is an extension of ethics' (Leopold 1949: vii–ix). How, we might wonder, are we to go beyond the social contract in the way hinted at by Leopold? Answering this question is one of the main objects of the second part of the present study.

2.2 Looking forward

As will become clear later, one trouble with the liberal, contractualist perspective is that it provides only a thin and rather
unbalanced account of human nature. My own account, however,
is not meant to compensate for this by giving a tremendously rich
characterisation of the human. In arguing that the modern
versions of the contractualist and utilitarian positions in moral
philosophy have ignored an important dimension of human
nature, I am simply engaged in moral criticism of the frameworks
they provide. The ecological contribution to the story of human
nature as I give it does not constitute, on its own, a rich account
of what we are. It leaves out many of the facets of our nature. My
central contention is simply that the ecological facet is of such
significance that we ought not to ignore it.

In failing to flesh out a rich account of human nature, I might
be accused of contributing one more 'thin' theory to the stock
already available. Such an accusation would involve
misunderstanding. The richness of human nature makes it
unlikely, on my account, that we will ever be able to depict its
details fully. Indeed, the idea of a complete description of human
nature, or of other natural phenomena, may be delusive. So I
have a conception of human nature as being extremely rich. If
what that nature is defies characterisation in any scheme or
framework, then all we can do, from time to time, is offer correc-
tives to those who concentrate excessively on one or two dimensions. We can also reflect that the project of morality will always
be incomplete. The very richness of our nature and character will
resist the attempt to impose tidy, often simplistic, theories on the
moral dimensions of our lives.

Some aspects of my strategy should now be getting clear. The
study of ecology is one branch of biology that furnishes a
framework within which we can think of ourselves and our relations to the rest of nature. As already suggested, this framework
is pleasantly compatible with others in the biological sciences and
has a role to play in informing our thinking on environmental
issues. At the same time, no single framework will provide the
definitive perspective on the problems to be faced in this work. For
these include the question of our relations with the rest of nature,
and involve — among others — moral, economic and political
assessments of our situation.

In order to find out what relevance *ecology* may have to our image of ourselves and of our relations to the things around us I begin the next chapter by distinguishing different things that are meant by those using the term 'ecology'. Certain positions concerning value in nature and dealing with the nature of human beings require us to believe that ecologists have uncovered important new truths about natural systems. In order to set these claims in an appropriate context, *scientific* ecology is distinguished from other kinds of 'ecology' (Chapters 3 and 4). Some popular appeals to ecological thinking are then seen to rest on one particular strand in the history of scientific ecology, a strand which is itself under sceptical scrutiny these days.

After this, it is necessary to look in particular at the question of whether ecological findings show that there is a special kind of 'holistic' and 'anti-reductionist' explanation. Much of the metaphysics that supports 'green' political and ethical thinking about the environment has been developed by somewhat mystical appeals to holism and a dismissal of reductionist explanations. I try to show that both of these metaphysical props are misconceived (Chapter 5). If 'green' thinking is to be supported by holistic, anti-reductionist metaphysics, this metaphysics will have to be maintained by appeal to familiar philosophical considerations; the science of ecology, as I understand it, gives no foundation for such a metaphysics.

What kind of explanation does ecology offer as a science? The deflationary answer is that it offers just the same range of explanations found elsewhere in the sciences — some causal, some of different sorts (Chapter 7). However, there is some confusion among writers as to whether stochastic, population-level explanations are in competition with causal explanations at a different level. By looking at some work on the theory of forest succession at the end of Chapter 7, I try to disentangle the issues — issues that are of relevance to the 'deep green' environmental thinkers who follow Naess in proclaiming themselves 'deep ecologists' (Naess 1973, 1984, 1986). The proposal that important ecological properties are supervenient (Chapter 8), overcomes problems in determining the functional role of whole natural objects.

The chapters in the second part of this book explore the political, social and ethical dimensions of environmental issues, on the basis of what I hope is a clear account of scientific ecology, ecological explanations and ecological entities. An immediate consequence is that much of the ground is cut from under the feet of those, like Callicott, Naess and Clark who would ask us to think about environmental problems in terms of a particular framework (Callicott 1985; Clark 1983). For that framework embodies a mistaken conception of ecology as a science and of the metaphysics that is associated with it. However, I sympathise with much of the ethical line propounded by these authors. In terms of the conception of moral philosophy as a *rationalising* enterprise, my aim is then to rationalise the sort of judgements that they and I make, without appeal to the models they use (Baier 1986). I do, however, deploy some modest ecological insights in the course of this rationalising enterprise.

In what has been written so far, I have been careful to adopt a studiedly neutral tone on the matter of whether an environmental ethic should be *biocentric, anthropocentric* or neither. If our ethical thinking is anthropocentric, we take human beings — or at least a significant number of them — as not only the beings who have moral concerns, and with whom such concerns originate, but also as the only proper objects of such concern. Now, however narrow an anthropocentric value perspective may seem to be, there are ways of getting other things in on the moral act while staying within the confines of that view. For since things other than human beings themselves matter to us, and are important to our welfare, due regard to human welfare will involve some consideration in respect of these other things. It is not difficult for an anthropocentric ethic to give value to clean water, fresh air, wholesome food, pleasant cities and decent working conditions.

If we move away from the human-centred ethical perspective, we can do so by giving some kind of *intrinsic*, or *non-instrumental* value to things other than human beings. The extension of ethical concern to animals in recent years is a good example of how this can be done. We can, of course, show concern and consideration for animal welfare out of a concern that is primarily focused on other humans. We may treat our domestic animals carefully so as not to upset the children or cause offence to the animal lover next door. But we can also be moved to show concern for the welfare of the animals themselves. That is, we can be moved to act in ways that respect the welfare, needs and interests of other creatures, even when we recognise that such creatures lack defining characteristics of humans. A biocentric ethic extends this kind of concern to all living things whatsoever, and has been widely supported in recent years (for example, Taylor 1986; Attfield 1983). Restrictions of the biocentric ethic may draw the line of concern around only a privileged class of living things, for example *sentient* or *intelligent* or obviously *conscious* things (for a review of such options, see Goodpaster 1978).

In terms of the extension of ethics mentioned at the end of the last section we might wonder if there are prospects for going beyond biocentrism. Not many thinkers these days take such a dramatic extension of ethics seriously, though I have previously tried to suggest that such an extension is intelligible (Brennan 1984). As I try to argue later in this book, standard biocentric treatments of the ethical leave aside much that is of serious concern to us. It is easy, of course, to define such concerns as lying outside the scope of morality, and I try to avoid getting entangled in terminological issues. Yet any reader of Leopold and Rolston is struck by the way their conception of an environmental ethic goes beyond the biocentric in one way, while being less than wholly biocentric in other ways (Leopold 1949; Rolston 1986). As I will argue later, recognition of what is called 'systemic value' in Rolston 1987 is compatible with taking a somewhat cavalier attitude to the welfare of some living things.

In the end, we have to recognise that there is not just one moral framework within which to articulate our thinking about the rights and wrongs of our dealings with nature. We can think about the same problem now anthropocentrically, and now in terms that are just as serious and compelling, although no longer anthropocentric. If the two perspectives give us the same result, our moral verdict is overdetermined. For virtually all the major environmental problems we face, such overdetermination is present — so the result of our deliberations is the same no matter which route we take. In the philosophically interesting cases where our deliberations arrive at different results, it may prove impossible for us, as human beings, to take seriously the judgement of the non-anthropocentric perspective. But that may be not so much a matter of morals but a reflection of what we are. Even if morality succeeds as a device for counteracting limited sympathies within the human community, it is unlikely to succeed as a device that will enable us to yield priority over human concerns and interests to the good of things 'natural, wild and free'.

3

Ecology: What It Is and What It Isn't

3.1 Two kinds of ecology

It is possible to distinguish at least two ways of thinking about ecology. On the one hand are the scientific studies of biologists, concerned with the interactions among organisms — whether taken individually or in groups — and between organisms and their environment. On the other, ecology can be regarded as a method of approaching problems, and as supplying a metaphysics that applies to far more than living systems. In this sense, ecology has application to academic disciplines and even to political and moral matters. Thus, it makes sense to think of some people as ecological psychologists, ecological physicists and ecological philosophers. Someone like Fritjof Capra, who claims that ecology has application to physics, is thinking of ecological methods as affecting the methodology of our approach to physics. We can label this non-biological notion of ecology *metaphysical* ecology and contrast it with *scientific* ecology.

In the scientific sense, there is an ambiguity of a harmless sort between ecology conceived in general terms as the study of organisms and their relations to other things, and the more specific study of the particular characteristics, life strategies, distribution and abundance of a certain kind of organism itself. Thus we can specify fields within ecology — like animal and plant ecology — or even think in terms of the ecology of field mice or slugs. In the specific sense of the term, scientific ecology is simply what embraces the study of the ecology of various organisms and kinds of organism. The support for their views claimed by metaphysical ecologists may be drawn from either the specific or the general studies.

31

In beginning to explore the relations between scientific ecology and ecology as a metaphysical stance let us start by looking at the sorts of things said by metaphysical ecologists. J.J. Gibson argues that the environment of a perceiving animal is different from the physical world. As he puts it:

> The mutuality of animal and environment is not implied by physics and the physical sciences. The basic concepts of space, time, matter and energy do not lead naturally to the organism-environment concept or to the concept of a species and its habitat . . . Every animal is . . . a perceiver *of* the environment and a behaver *in* the environment. But this is not to say it perceives the world of physics . . . (Gibson 1979: 8)

Is there more to the ecological approach to perception than the claim that animals and their environment stand in mutual relations — so that we cannot have one without the other? Gibson goes on to suggest a number of other ecological doctrines. First, there is the nesting of smaller units of the environment in other, larger units. Thus boulders and soil are nested within canyons, which are in their turn nested within mountains. This nesting is not hierarchical, but involves interesting overlaps. Thus there is no preferred metric, no one scale or set of fundamental units, in terms of which the environment can be described. 'The unit you choose for describing the environment,' he writes, 'depends on the level of the environment you choose to describe' (Gibson 1979: 9). In particular, then, there is no privileged or fundamental description of the environment possible in terms of elementary particles, or in terms of chemical structure.

Gibson clearly thinks these distinctions lead to important differences between ecological and physical descriptions of the world. Think of what happens when a cube of ice melts, a liquid evaporates or an object is destroyed by fire. In terms of physics, we know that there is persistence through such changes. But there is no persistence from an ecological viewpoint: 'Ecology calls this a *non-persistence*, a destruction of the object, whereas physics calls it a mere change of state. Both descriptions are correct, but the former is more relevant to the behaviour of animals and children' (Gibson 1979: 13–14). The claim is, then, that the framework of physical events is different from the framework of ecological events, and we cannot describe the latter purely in terms of the former.

There are two more aspects of Gibson's metaphysical ecology worth noting at the moment. First is the claim that just as there are physical laws, there are ecological laws. Gibson distinguishes in the perceiver's environment among the medium (something like air or water which permits motion and transmits light), substances (things which prevent locomotion and do not freely transmit light or odour) and surfaces (which separate the medium from the substances of the environment). Notice that these distinctions are relative: water is, by and large, a substance for us (and thus has a surface), while it is a medium for fish.

In terms of these distinctions, Gibson is able to put forward certain suggested laws. For example, one ecological law he gives is that 'any surface has a characteristic texture, depending on the composition of the substance whose surface it is (Gibson 1979: 24). In effect, this 'law' suggests that rock, clay, wood, ice, sand and other substances will each have distinctive surface textures. Surface texture is an important feature in our relation to our environment. Without it, we would fail to recognise the differences between paper and wood, cloth and concrete.

As well as instancing such 'ecological laws', Gibson maintains that the distinction between perceiver and the environment is not fixed by the surface of the skin. The capacity to attach things to the body — in particular tools — shows that, as he puts it, 'the absolute duality between "objective" and "subjective" is false' (Gibson 1979: 41).

Like Gibson, Capra denies that there is any fundamentally privileged description of the world's events to be given in terms of physics or the other 'hard' sciences. He writes:

This world view of modern physics is a systems view, and it is consistent with the systems approaches that are now emerging in other fields, although the phenomena studied by these disciplines are generally of a different nature and require different concepts. In transcending the metaphor of the world as a machine, we also have to abandon the idea of physics as the basis of all science. According to the bootstrap or systems view of the world, different but mutually consistent concepts may be used to describe different aspects and levels of reality, without the need to reduce the phenomena of any level to those of another. (Capra 1983: 89)

Ecology is supposed to typify the systems view of reality, of which Capra here writes; according to him, it takes a holistic, anti-reductionist approach. Since the terms 'holism' and 'reductionism' are particularly troublesome and lack clear meaning, a later chapter will give attention to their clarification. For the moment, observe that Gibson and Capra also agree on the fact that there are interesting ecological — or systems — laws, and that there is no absolute distinction between subjective and objective.

Just as Gibson denies the existence of a fixed hierarchy among the levels in natural systems, so does Capra: '. . . most living systems exhibit multi-levelled patterns of organisation characterised by many intricate and non-linear pathways along which signals of information and transaction propagate between all levels, ascending as well as descending'. Capra continues, on the same page, to make the following striking claim:

> The various systems levels are stable levels of differing complexities, and this makes it possible to use different descriptions for each level. However, as Weiss has pointed out, any 'level' under consideration is really the level of the observer's attention. The new insight of subatomic physics also seems to hold for the study of living matter: the observed patterns of matter are reflections of patterns of mind. (Capra 1983: 305)

Although the example of tool manipulation makes plausible the claim that there is only a relative distinction between what is subjective and what is objective, Capra is apparently making a much stronger claim than this. For he seems to think that ecological, or systems, descriptions will lead us into a kind of *idealism*, using that term in its philosophical sense. Metaphysical ecology is therefore heady stuff: it threatens to bring back the idealist doctrine that the features we discern in the world around us are in some real way *mind-dependent*.

The striking similarities between Gibson and Capra suggest that the study of ecology is capable of yielding ideas and a methodology that take us far beyond the scope of biology. In a moment, we will consider whether the kind of view urged on us by these writers is justified by the scientific work that has been done in ecology. In the end, however, such an inquiry raises the question of the scientific status of ecological investigations themselves.

It is interesting to conclude this brief review of metaphysical ecology by looking at the extreme claims made recently by one philosopher. Encouraged by Capra's dismissal of the object–subject distinction, and worried by the question of what gives nature its value, Baird Callicott writes:

> . . .the central axiological problem of environmental ethics, the problem of intrinsic value in nature, may be directly and simply solved. If quantum theory and ecology both imply in structurally similar ways in both the physical and organic domains of nature the continuity of self and nature, and if the self is intrinsically valuable, then nature is intrinsically valuable. If it is rational for me to act in my own best interest, and I and nature are one, then it is rational for me to act in the best interests of nature. (Callicott 1985: 275)

To be fair to Callicott, it must be observed that he is simply following up the implications of taking Capra's ecological holism seriously. Yet, if contemporary ecology gives any support to the kind of metaphysic lying behind Callicott's claims, it is important that we recognise this without delay. So let us now look at scientific ecology, and at the actual theory and practice of some ecologists.

3.2 Scientific ecology

The definition of a science that is still finding its feet can only be provisional. A popular account of what ecology is suggests that it is the kind of biology that deals with organisms from the skin out, as opposed to studies that look under the skin. However, this will not do as a serious account of the subject. For a start, many contemporary ecologists spend much of their time looking at the internal details of organisms, and not just because what goes on inside an organism is relevant in giving an account of its relations to other things. On the contrary, there is scope for ecological study within the gut of a ruminant, for example, just as within the margins of a pond.

A suitably broad definition of scientific ecology is to take it as *the study of those interactions with their environment that determine characteristics, distribution and abundance of organisms and systems of organisms.* Such a definition immediately requires qualification. Not all

35

characteristics of organisms are determined by their interactions with other things: the genetic constitution of a particular creature, for example, may determine its shape, or the size of its eyes, irrespective of variations in its environment. Moreover, 'systems' as used in the definition above, does not refer to the technical concept used in systems theory: rather, it is a vague term covering populations, species and communities of organisms however structured.

With these qualifications, however, the definition captures some of the features implicit in Haeckel's original definition (in 1866) of ecology as the study of the interactions between organisms and environment. Some ecologists would insist that only distribution and abundance should be mentioned; but such an insistence flies in the face of actual practice where properties of organisms or communities other than those relating to abundance or distribution are carefully monitored. Thus, in a series of studies, one ecologist studied the size and larval period of various amphibians (Wilbur 1984). Of course, the results of the study can be presented in terms of the distribution of body sizes of tadpoles, for example, in certain conditions. But the work also gives a means for answering such questions as: 'why is this tadpole bigger than others?' Interactions among organisms can determine individual characteristics as well as populational properties of distribution and abundance.

Notice that I am not arguing that ecology is properly concerned with individuals, or with unique single cases. As with any other branch of the sciences, ecology is concerned with individual items to the extent that their behaviour and properties exemplify general truths or laws. As we will see, not all scientific laws are causal generalisations. But consider, for the moment, the case of such generalisations. Suppose that competition from an introduced species causes an existing population to switch to different food resources from the ones previously used. Ecologists are reasonably good at predicting just which introduced species in such situations will bring about a change in the behaviour of established species in an ecosystem or community. They are thus able to make causal generalisations; but we can explain individual behaviour precisely by appeal to the general principle (for more on laws in biology see Chapter 3 of Hull 1974).

Since ecology is still very much in its infancy, only a limited amount of agreement is possible when we turn to questions of definition. Generally, the vaguer and less helpful the definition, the more likely there is to be widespread agreement about it.

Worse, since ecology overlaps with other areas of biology, there is the possibility of disputes over what ecology is which are partly terminological and not easy to resolve given the present state of the subject. Thus population genetics and evolutionary theory seem to have much in common with ecology. Both, for example, are concerned with properties of populations, and the relation of one population to others, and to features of the abiotic environment. One simple way of sorting out the difference between ecology and evolution theory is to think of ecology as concerned with organisms, or populations, over short stretches of time. Whether a fish will grow to a certain size depends, to some extent, on the availability of food resources for it, its competition for these resources with other fish, and perhaps also on features pertaining to water temperature, pollution and so forth. These ecological constraints on its growth operate after its arrival in the environment being studied. But selective pressures over thousands of years have equipped that fish, and others like it, with a genetic program which itself produces its response to the environment. Those aspects of its program which the organism can pass on to its descendants, the heritable features it possesses, will themselves vary in their distribution over fish populations through time.

Now, this change in heritable properties over time is evolution: and it is generally thought to be in turn influenced by environmental pressure over time. So now we can distinguish two kinds of pressure produced by an environment. First, there is the pressure which determines that a species population of fish, equipped with a certain genetic program, will grow to a certain range of sizes — perhaps the range that is optimum for the environment in which it is placed. These pressures will be the sort the ecologist studies. By contrast, there is pressure exerted through long periods of time by a perhaps changing environment which leads to a different distribution of heritable characteristics across a species population. This changing distribution of characteristics can be explained by differential reproductive success on the part of different organisms. A property which confers high reproductive success on an individual in a certain environment and which is inheritable by the offspring of that same individual is likely as time goes by to become more widely shared among members of the population occupying the same environment. The study of these pressures operating through time is the province of evolutionary biology.

Viewed in these simple terms, evolutionary biology and

ecology are complementary. The one gives an account of competition for resources, and distribution of characteristics over relatively short time periods. The other seeks to explain the changing nature of populations and species through long time periods. Only by combining both sorts of study can we hope to get satisfying answers to questions of why organisms in certain situations have the properties they display instead of quite different ones: the satisfying answer will tell us that it is partly a matter of interaction with other things and partly a matter of inherited programs. In the end, our ignorance of just what particular route the various pressures have forged for the history of species means that we may be able only to speculate about the actual history of life on earth. We need to be suspicious of 'just-so' stories told by theorists — a suspicion that does not cast doubt on the standing of evolutionary theory itself.

3.3 Ecologists at work

Here is an almost random selection of recent work, showing the range of contemporary research in scientific ecology. The relation of prey to predator is a longstanding interest. Think of the physical constraints on predation imposed by the simplest characteristics of both predator and prey. For example, if a creature takes its prey whole, then the size of the prey and the size of the predator's mouth parts are going to be correlated. In a recent study on fruit-eating birds, one ecologist looked at the relation of a bird's gape width to the preferred fruit size (Wheelwright 1985). Dealing with 70 different bird species and more than 160 species of plant in the lower montane, or hillside, forests of Monteverde over five years, the study showed that although the gape width puts an upper limit on size of fruit taken, there was no correlation between the smallness of fruit taken and the gape width of the bird species studied. Indeed, it turns out that many broad-gaped birds eat widely of small fruit. Looked at from the point of view of the plants, so to speak, those offering the largest fruit were relying for seed dispersal on birds with broad gapes; yet such birds were general feeders, not themselves specialising on those plant species with the largest fruit.

Staying with the business of swallowing and physical constraints associated with that, another recent study on one species of lizards shows that the energy they expend catching and eating crickets is only a minute fraction of the utilisable energy of the crickets (Pough and Andrews 1985). In other words, the energy costs to lizards involved in subduing and swallowing crickets can be ignored for most purposes. This research, like the previous piece, is of relevance in testing certain widely canvassed mathematical models dealing with predator–prey relations. For instance, it might be held that an optimal diet for either a fruit-eating bird or a cricket-consuming lizard is one where a maximum of energy is gained for the amount of time spent feeding. Until recently, it was common for mathematical models to be worked out for optimality in such cases, with only a minimum of observational evidence to support the model. Trends in recent ecology are to move to a more experimental, observation-based approach.

Each of the studies mentioned above is essentially concerned with resource utilisation. However, if resources are at all scarce, there is likely to be competition among species in a given area for the resources boasted by that area. Some species, although quite different from each other, seem fitted for using the same resources. Where such species occupy territory exclusively, with no evidence of other species who might compete for the same resources, they are said to occur *allopatrically*. By contrast, so-called *sympatric* species are potential competitors for the same range of resources and occur together in the same physical areas. The thought that distribution and abundance of species in habitats reflects interspecific competition for resources is one that is so natural that it has enjoyed a central place in ecological thinking. Recent work has set about testing certain intuitively plausible notions about the nature of such competition.

A recent study of sympatric hummingbirds in Arizona seems to show that interspecific competition can lead to a dominant, more aggressive population gaining access to better food resources than the less aggressive species populations occupying the same territory. This study shows great ingenuity in the construction of tests of a certain model of interspecific competition (Pimm, Rosenzweig and Mitchell 1985). The researchers identified three species of hummingbird, two of which showed no interactive effects on one another. The third species, however, the blue-throated hummingbird, dominated the other two. The researchers were able to control the quality of food offered to the hummingbirds in a certain area (by offering weaker and stronger sugar solutions) and were also able to study the changing densities of the dominant species. The results of the study were quite startling.

Not only did all species of hummingbird prefer the stronger sugar solutions when other things were equal, but as the density of the blue-throated hummingbirds increased, the result was not simply that all the species were forced to feed from both richer and weaker solutions. Rather, the blue-throated hummingbirds made more use of the weaker solutions as the density of their own species increased. At maximum density, the blue-throateds were spending only 40 per cent of their time at the rich feeders. However, as the blue-throateds increased in density, the other two species became more and more dependent on the weaker solutions. At maximum blue-throated density, the other two species were exclusively using the feeders containing weaker solutions.

Is not all this just what we might expect? As the authors of this particular study point out, there are other outcomes that could have come about. One *null hypothesis* is that feeding behaviour is not affected at all by the density of the hummingbird population in the area under study. A biological null hypothesis is simply the hypothesis that nothing has occurred warranting explanation in biological terms. But the observations recorded make clear that feeding behaviour does change with increasing density of the hummingbirds. Another hypothesis is simply that birds will switch under the pressure of increasing density to use all the available sources of energy. But the experiment shows that this hypothesis holds only for the dominant species. It is not true of the two subordinate species which, under pressure from the increasing population of blue-throateds, switch their feeding behaviour so that they 'specialise' in the alternative, poorer energy source.

Thus the experiment shows that something worthy of explanation has occurred, and the researchers claim additionally that their observations confirm a certain graphical model of interspecific competition. In the conclusion to their paper, they argue that, over time, it is not impossible that natural selection might so operate as to yield species of hummingbird that exclusively use the separate resources. In this way, ecology and natural selection would yield an explanation for the differential behaviour of similar species in the same natural system: initial competitive exclusion might lead in the end to non-competitive foraging behaviour. This work is thus a contribution to the long-standing debate on the issue of niche-differentiation.

Clearly, the conclusion is speculative. Yet the research illustrates nicely the point made provisionally in the previous section about the complementarity existing between explanations at the level of evolution and those at the level of ecology. Moreover, the study of competition for resources brings us to something of central concern to students of ecology since Haeckel first coined the term. With its interest in graphic models for competition for resources, ecology bears more than a passing resemblance to economics: this observation prompts a further convenient definition of the subject: ecology is the systematic study of the economics of nature.

Perhaps one of the most fascinating things about studying any science is the way that discoveries and observations at one place or time echo those at another. Thus, the findings about humming-birds correlate nicely with some well known results from the history of the subject, only these results deal with very different kinds of organisms. Studying the microscopic protozoan *Paramecium*, the Russian ecologist Gause found that competitive exclusion apparently worked for them. Having grown bacteria and yeast cells which fed on oatmeal in a liquid medium, Gause introduced three species of *Paramecium*, each of which was able to thrive alone on the bacteria in the medium. However, in test tubes in which a specific pair of protozoans were grown, one would regularly thrive while the other declined to extinction. The 'dominant' protozoa — *Paramecium aurelia* — was, however, not lethal when grown together with the third. In this case, the two species of *Paramecium* lived together in the same tubes, but there was a partitioning of resources: *P. aurelia* lived mainly on the bacteria suspended in the medium, while the other species lived on the yeast cells at the bottom of the tubes.

Gause's experiments were carried out before the second world war. Yet the kind of competitive exclusion they apparently reveal has been observed in many other experiments before and since (Begon, Harper, Townsend 1986, Ch. 7.2). A human ecologist would not find it surprising if intraspecific competition among humans showed some of the same features — for instance, a parti-tioning of resources so that those less dominant survive on a perhaps poorer resource than that utilised by the more dominant. However, whether with humans or with protozoa, we have to be clear in our understanding of notions like *competition* and *dominance*, something that is not easy when they are applied in such diverse cases.

3.4 Null hypotheses

Let us stop to think in more detail about the use of null hypotheses, as illustrated in the case of the hummingbirds. Great controversy in recent ecology has focused on the issue of whether certain phenomena need to be explained in biological terms at all. Consider, for example, the issue of species distribution on islands — not just physically separate islands, but any clump of resources remote enough from other clumps to make their associated communities relatively isolated from other communities. Here is a plausible notion of what happens in such isolated assemblages. If there is competition for resources among sympatric species, then where further resources are easily available, the less dominant species can retreat so that they are able to avoid confrontation with the dominant species. Thus, in habitats that are not isolated from others, we will tend to find rich representations of species. Think, in this context, of the amazing richness of tropical rain forests in terms of the number of species present in any given area.

By contrast, we might reason, where a clump of resources is isolated from other clumps of resources, then those dominant species that happen to be established on that clump will be able to secure the best of the resources for themselves. Weaker species will tend to obtain very little by way of resources, and will thus die out. So an ecological island is liable to show a poorer representation of species than mainland communities show. We can try to put this idea into terms that are easily quantified. Take, for example, the ratio of species to genera in an area. Elton pointed out in 1946 that large regional composites of plants will have around 50 per cent of plant genera represented by only one species. Switch attention to islands, and we find that around 84 per cent of genera have only one species represented. The richness of large communities seems therefore beyond doubt.

Although Elton thought these findings were a clear vindication of the theory of competitive exclusion, he had apparently not considered the appropriate null hypothesis. As Williams later showed (see Williams 1964), islands are in fact rather richer than might be expected given the mathematics of random aggregation (see further Järvinen 1982). Suppose, for example, we take the suits in a deck of cards to represent four genera of plants, with each card representing a species. If we deal out 20 cards, the likelihood is that each suit will be represented by more than one card. This likelihood decreases, however, as we deal out smaller and smaller numbers of cards. If we deal out only seven cards, it is not unlikely that a certain suit is represented by only one or two cards, and sometimes by none at all. The interesting thing about islands, viewed from this perspective, is that their species-to-genus ratios are so high. And that means, as Strong puts it: 'any net forces influencing the richness of the taxonomic hierarchy in these real communities must be the opposite of competition' (Strong 1980: 275).

The importance of null hypotheses, and the whole issue of randomness in ecology have been brought to prominence in recent years through the work of Strong, Simberloff and others. Not surprisingly, some students of island ecology, most notably Diamond, have tried to show that competitive exclusion, and other features, can explain the non-random distribution of species on islands and archipelagos. Although the issue of null hypotheses will recur later it is worth noting some initial difficulty with the issue as posed so far. Although there may be nothing to explain in the discovery that a set of seven cards has two clubs, three diamonds, one spade and one heart (supposing the hand had been randomly dealt from a well shuffled pack), the same result could have been carefully reached by someone selecting the cards according to a non-arbitrary rule. In other words, the fact that a certain distribution of species might have come about by chance does not show that it in fact did so. This difficulty is one that confronts both ecology and the theory of evolution. It may be that a certain animal is adapted by natural selection to fulfil a certain role or occupy a certain niche; but that may not be the reason why that animal is found in the place it happens to occupy in the system under study. That natural selection is a possible explanation of a phenomenon in no way guarantees that the phenomenon had to occur, or did occur, through the process of selection.

In the fascinating field of *biogeography*, the status of our current theories about community history are very much a matter of controversy (Strong, Simberloff, Abele and Thistle 1984). In some sense, ecological communities can be thought of as complexes of 'islands' in that species sometimes display distinctive ranges. Likewise, diversity of species — a feature that is highly pronounced in tropical forests — might be thought to be explicable by references to past disruptions in climate or habitat, rather than due to chance factors (see Beven, Connor and Beven 1984 for interesting accounts of this issue). Even for species–area distributions (where it can be observed that there is a positive relationship between the number of species found in an area and the size of that area) there is no clear agreement on the operation of chance factors as against biologically significant factors concerning habitats or patterns of disturbance (McGuinness 1984).

It is tempting to compare evolutionary ecology — the matching of population genetics and community ecology — to cosmology. Just as the physicist can trace the history of the solar system to some primordial state, and trace the entire universe back to the big bang, so, we might think, the ecological historian can trace the origin of individual species and communities to their early beginnings. But the very complexity of biological phenomena threatens this prospect. We are not dealing with items reducible to simple particles with simple properties. Rather, the working of ecological rules and principles may involve the interplay of chance and coincidence along with plausible principles of structure and organisation. If this is so, then probabilities are about as good as we are likely to get in any historical forays we make on the biological front. The power of sociobiology to give explanations of human nature and conduct is thus severely limited — although, as already noted, sociobiologists are perfectly able to make up appropriate 'just-so' stories about the origins of our drives, ambitions and characters.

To the extent that ecology is concerned with different levels of organisation, it is the level of community organisation that seems most significant for its application to other areas. The systems view of life which Capra dwells upon is a view influenced by theories of community structure. Competition and mutual co-operation are the two obvious regulators of such structure. If, however, the suggestion that communities are not structured according to any principles other than those governing random walks or distributions of playing cards is correct, a great many people have been misled about the nature of ecological insights, and the prospects for metaphysical ecology in this respect at least become dramatically poorer.

4

Ecology in Perspective

4.1 Some history

Ecology is concerned with species populations as members of communities inhabiting a certain abiotic background in which there may be competition or co-operation. The small number of examples so far given could be multiplied. But to do so would reinforce an impression that should already be obvious. The way organisms fare in what Darwin called the 'struggle for existence' will depend on their own characteristics, the characteristics of other organisms around them and the nature of the abiotic context. At levels of organisation above the population, there may be principles of community structure which explain the distribution of species in a certain area, and the nature of the niches they occupy. One aspect of contemporary ecology not mentioned so far concerns the interactions among very different kinds of species. For example, animal species are on the whole more mobile than plant species. Because bees fly around moderately freely, while the plants they pollinate and feed on are static, we tend to think of bees as utilising plants as resources. But, equally reasonable on a resource model is the notion that bees are resources (pollinators) for plants. So, for example, we could look to see whether certain plant assemblages are organised in such a way as to reduce competition for pollinator service (Armbruster 1986).

The examples mentioned so far have involved little dependence on such well known ecological concepts as those of *niche*, *succession*, and *food chain* or *food web*. These concepts themselves developed as ecology developed from its origins in the study of natural history. Indeed, in 1927 Elton defined ecology as no more than scientific natural history. Nowadays, ecologists would be uncomfortable

45

about accepting such a definition, for it suggests something at once too amateurish and too descriptive. If we go back to the eighteenth century, the age of reason, we can note that the idea of a structured universe and a rational nature seemed attractive to natural historians. In a much quoted passage, Richard Bradley wrote in 1721, 'All Bodies have some Dependence upon one another; and . . . every distinct Part of Nature's works is necessary for the Support of the rest; and . . . if any one was wanting all the rest must be consequently out of Order' (McIntosh 1985: 70). The subsequent work by Malthus on population growth as a function of resources and food supply not only provided Darwin with ideas but very likely contributed to ecological thinking. Certainly, by the end of the nineteenth century, it was widely held that nature displayed an equilibrium or balance within which species functioned as orderly, integrated units.

Another theme of the nineteenth century was the notion that the ordered balance of nature was only disturbed by human intervention:

> There is a general consent that primeval nature, as in the uninhabited forest or the untilled plain, presents a settled harmony of interaction among organic groups which is in strong contrast with the many serious maladjustments of plants and animals found in countries occupied by man.
> (Forbes 1880, quoted in McIntosh 1985: 74)

It was Forbes who also first suggested that groups of plants are like superorganisms, massive individuals with their own principles of organisation and flourishing: 'A group or association of animals or plants is like a single organism in the fact that it brings to bear upon the outer world only the surplus of forces remaining after all conflicts interior to itself have been adjusted' (Forbes, quoted in McIntosh 1985: 75). A few years later, Clements gave a classic statement of the same view:

> The plant formation is an organic unit . . . According to this point of view, the formation is a complex organism, which possesses functions and structures, and passes through a cycle of development similar to that of a plant . . . Since the formation, like the plant, is subject to changes caused by the habitat, and since these changes are recorded in its structure, it is evident that the terms, function and structure

are as applicable to the one as to the other. (Clements 1905: 199)

Assessing this view nearly 50 years later, Gleason wrote:

There is nothing comparable to reproduction in any assemblage of plants . . . Far from being an organism, an association is merely the fortuitous juxtaposition of plants. What plants? Those that can live together under the physical environment and under their interlocking spheres of influence and which are already located within migrating distance. (Gleason 1952: 8–10)

Clements is not only notorious for championing the superorganism conception of the plant association. He also introduced the notion of *succession* into ecology. This useful notion is one of the bugbears of contemporary ecology. Just as people seem to divide largely into dog-lovers and cat-lovers, so ecologists seem to divide into two classes: those who think there is something in the theory of succession, and those who do not.

The basic idea is entirely natural. Suppose we start by considering a hitherto uncolonised, bare area. Perhaps it is at the margin of a lake. At first, it is colonised by various kinds of water plants whose roots trap and accumulate silt and soil. Over time, the area thus colonised begins to dry out a little and become marshy. New species now colonise the marsh, while the original occupants of the area flourish only along the margins of the now shrinking lake. Some hazel and alder trees, let us imagine, take root in the marsh, and with their large root systems and heavy uptake of water they gradually change the nature of their surroundings still further. As succession continues, the alders and hazels give way to broad-leaved woodland, which is the *climax* to this particular successional process. Clements, in fact, thought that succession led on the whole to very limited outcomes: 'Grassland or forest', he writes, 'is the usual terminus of a succession' (Clements 1905: 265).

Notice that, whatever its merits or demerits, Clements' theory tries to do justice to the dynamics of change in nature. Indeed, he thought there were principles or laws of change, although his formulation of these is somewhat vague. One famous principle, which he called the 'law of reaction', states:

Each stage reacts upon the habitat in such a way as to produce physical conditions more or less unfavourable to its permanence, but advantageous to the invaders of the next stage. (Clements 1905: 265)

Clements' plant communities, incidentally, are not themselves *ecosystems*, although he anticipates later work on such systems. In the sense first defined by Tansley in 1935, an ecosystem contains populations of both plants and animals together with a number of abiotic materials which are cycled through the system (for example, solar energy, carbon, nitrogen and water). Communities, then, would be ingredients in ecosystems, and study of the dynamics of communities would be a contribution to the study of ecosystem dynamics.

We can take this disagreement between Clements and Gleason as typifying the whole issue of the status of ecosystems. We can distinguish two extreme views here. Such systems are either, in Tansley's phrase, 'the basic units of nature' (Tansley 1935), items of enormous complexity of structure within which we, and all other living things are nested, or else we wishfully impose pattern and structure where there is merely chance aggregation.

With all this interest in community dynamics, it is not surprising that individual organisms and populations of organisms were thought to have a place and function in such large communities. In 1927 Elton described the places of populations of animals in terms of four factors. One was the notion of a food chain or a food web; another was the idea of food size. A third mode of assigning places to populations was by means of the 'pyramid of numbers', the idea being that large numbers of smaller creatures are required to sustain a small number of larger creatures; thus increasing size is correlated with reducing numbers, and populations of large animals increase much more slowly than populations of smaller ones. In modern terms, we would say that in grazer communities, the secondary productivity of herbivores is much lower than the primary productivity of the autotrophs, though this generalisation does not hold good for other kinds of community (for example, tree- or plankton-based ones). The final factor, however, was the central concept of the *niche*. This was not just the physical location of an animal: that is, its habitat. Rather, for Elton an animal's niche was something like its 'status in the community' or 'its relation to food and enemies'.

On this conception of the niche, more than one species could fill the same niche, for it was rather like a functional role that species had in the broader network of populations constituting its local community. Although earlier accounts of the niche had suggested that filling a niche was the result of interspecific competition (see Grinnell 1908), Elton was unimpressed with competition in his early work on animal ecology. Thus a species' Eltonian status in a community of other species need not depend on a history of past competition, although the case of the hummingbirds described in the last section would suggest that such competition may sometimes be relevant.

Elton's somewhat loose definition of the niche contrasts interestingly with later accounts. In an important landmark in the history of what is sometimes wrongly called 'new' (in the sense of mathematical) ecology, Hutchinson gave a geometric definition of the niche. Suppose, for simplicity, we consider a type of organism for which temperature is absolutely critical: it is a plant, let us suppose, which cannot survive if the ambient temperature fluctuates more than a certain amount over the day. A certain range of temperatures then constitutes this plant's niche in one dimension. However, temperature may not be the only critical thing. Our imaginary plant species is also very sensitive to humidity, to salinity, to the level of the water table and so on. Each of these conditions adds a further dimension to our description of the plant's niche. Now imagine the case where there are n such conditions. Just as we can map three conditions by a graph or diagram in three dimensions, so we can conceive of the n conditions as defining a volume in n-dimensional space. And this is precisely what Hutchinson's definition states (Hutchinson 1957). The ecological niche of a species is the volume in n-dimensional hyperspace within which it can maintain a viable population.

We can use Hutchinson's definition to draw attention to an interesting problem in niche theory. Consider, for a moment, all the environmental conditions — including abiotic mineral and energy resources — that might conceivably be of relevance to the maintenance of a species. Clearly, the list is likely to be enormous. Yet, in a real situation, there will be constraints on the access a species has to these various resources, since many of the same resources will be utilisable by other species. Even sunlight is something that can be denied to one plant by the presence of another. Hutchinson was therefore forced to distinguish between the *fundamental* (theoretical) niche of an organism and its *realised* niche. The realised niche will occupy a considerably smaller volume in hyperspace than the theoretical, fundamental one. The same species may also contain populations in different environments occupying quite different realised niches.

Since the same species, when represented in different habitats, may not display the same realised niche, it will not be easy to tell what constrains the realisation of the niche in a given set of circumstances. Think again of Gause's protozoa experiments described in the last chapter. There was no difficulty about finding the realised niche of the three species under laboratory conditions. Moreover, it seems clear that one species of protozoa was driven to extinction when its habitat included one of the others. In this case, the species driven to extinction lacked a realised niche when in competition with the stronger species. Yet what about the case of the apparently competing species which found a way of partitioning resources? Here, both species had realised niches, but they were distinctly different. Just as extinction seems associated with competition, so the difference of realised niches also seems to be, and indeed, Gause's principle of competitive exclusion — one of the traditional 'laws' of ecology — maintains: *when two competing species coexist in a stable environment, then they do so as a result of niche differentiation; if the habitat precludes niche differentiation, however, then one of the competitors will eliminate the other.*

This association of interspecific competition with differentiation of realised niche may look very neat. Yet there are problems about testing the claim. For now suppose we look at the occupants of a real biological system. The laboratory manipulations, like the manipulation of the hummingbirds by Pimm and his associates, may show us examples of current competition associated with niche differentiation. But that does not prove that new cases observed in the field where similar species coexist but differ slightly in feeding preferences are examples of current competition. It may be that, as a result of past competition, for example, those members of one of the species that preferred to feed rather than argue will have had more offspring. Natural selection will thus have favoured those who have avoided competition. If this were so, the removal of one species would not have any effect on the distribution and abundance of others. By contrast, a further explanation does not involve competition at all. On this theory, we imagine the coexisting species as having evolved simply to be different. In the past, some species had one feature, preferring food of a certain size and flavour, and other species differed slightly. They still do. But they never competed in the past, and they still do not. Studies of protozoa or of hummingbirds, then, have only a limited relevance in confirming generalisations about competition and exclusion. Just what the scientific ecologist can do in the face of this sort of problem is best tackled later when we look in more detail at the relations between theory and evidence.

This detour into niche theory was prompted by considering the mathematical definition of the concept of the niche. It is unfair to characterise mathematical ecology as being new, as if early ecologists were not themselves interested in mathematical modelling or in formulating laws in terms of equations. The difference between the period before Elton and the period after Hutchinson is one of degree, not of kind. Many modern treatments of ecology emphasise the use of mathematical and stochastic models to a far greater degree than was the case earlier in the century. It is worth noting that the history of mathematical ecology goes back certainly to the early nineteenth century, when principles of population growth were formulated as differential equations relating change in numbers of a population over time to the ceiling population, or carrying capacity of that population's environment (McIntosh 1985: Ch. 5). In this century, Pearl, his student Lotka and the physicist Volterra used the same method of differential equations to represent single and multi-species situations. The so-called Lotka-Volterra equations have as honourable a place in the lore of ecologists as Gause's principle. The graphs represented by the Lotka-Volterra equations for interspecific competition bear an uncanny similarity to the graphs of the results of Gause's experiments with *Paramecium* (see Chapter 4.2).

It may seem odd to have gone thus far in sketching some of the history of ecology without having said much about ecosystems themselves, nor about the notion of a food chain or web. In looking at these notions we encounter what was, for many, an exciting 'new' development in ecology. Indeed, for some outsiders, ecology is a paradigm of a systems science — as we have already seen. Yet although holistic systems notions have played their part in the thinking of naturalists and ecologists, not all ecologists have been equally attracted to them. The same kind of split apparent nowadays between the holists and the anti-holists was present in the past. Some writers have even suggested that plant ecologists were traditionally more inclined to systems thinking than animal ecologists (McIntosh 1985: 169–71). One reason for this may have been the obvious tendency for plants to clump together into associations, formations or communities, and the relative ease of sampling many species at a time. Animal sampling, always a much trickier business, perhaps encouraged animal ecologists to study populations separately.

After Tansley's first introduction of the term 'ecosystem', it was mainly used in connection with the description of aquatic systems, and after Lindeman's classic paper on tropho-dynamics of 1942, it was still limnology and aquatic biology that provided the major source of data for ecosystems theory. What Lindeman's paper did was bring together a number of trends, producing a theory that made a systems view plausible, which built on Elton's earlier insights (the insights, ironically enough, of an animal ecologist) and which teamed up synergistically with the theory of succession. Elton's pyramid of numbers now obtained a new, and exciting, application, being transformed in the process into a pyramid of energy and matter. Within the ecosystem, energy passes upwards through a number of what Lindeman called *trophic* levels. The passage of energy and matter constitutes a food cycle, and the organisms at different trophic levels constitute links in a food chain, or — perhaps a more accurate metaphor for what Lindeman had in mind — as knots in a food web (Fig. 4.1).

At the heart of the ecosystem, thus conceived, lie the *autotrophs* — organisms that can transform inorganic material into organic compounds. Without them and the decomposers, which provide the essential recycling mechanism, solar energy could not be used to support life within the system at any level of complexity. *Heterotrophs* rely for their energy needs either on consuming autotrophs, or on consuming other heterotrophs. Autotrophs occupy a lower trophic level than heterotrophs. Location in a food web is also associated with other factors — like size, efficiency in use of food, loss of energy due to respiration and so on. Once we see items as linked elements of a web, the web itself becomes a natural unit of study — although it is worth noting that by and large ponds and lakes have so far been the most plausible examples of ecosystems (but see Golley 1960 for plausible models of terrestrial ecosystems). Lindeman's significant advance over the notion of previous plant ecologists that plants constituted communities, or formations, is his explicit recognition of the role of the abiotic — minerals, solar energy, water and various elements — in supporting, and interacting with, living communities. As he put it: '. . . analyses of food-cycle relationships indicate that a biotic community cannot be clearly differentiated from its abiotic environment; the ecosystem is hence regarded as the more

Ecology in Perspective

Figure 4.1: Model of trophic structure

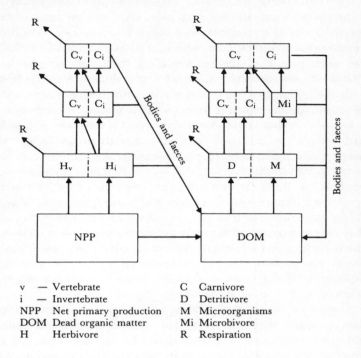

v — Vertebrate
i — Invertebrate
NPP Net primary production
DOM Dead organic matter
H Herbivore

C Carnivore
D Detritivore
M Microorganisms
Mi Microbivore
R Respiration

Source: Begon, Harper and Townsend 1986, after Heal and MacLean 1975

fundamental ecological unit' (Lindeman 1942: 415).

Although modern research does not validate all the claims made about ecosystem development, it did look for a while as if the theory of tropho-dynamics and the theory of succession could be integrated into a comprehensive view. On this view, ecosystems are dynamic wholes which grow to maturity by conserving biomass, increasing diversity of species and evolving mechanisms to damp down oscillations and eventually form states of *stability* (although such states would not be, in thermodynamic or chemical terms, states of high *equilibrium*). This stability and complexity of organisation is maintained despite continual immigration and emigration of species. The combination of tropho-dynamics and succession theory thus made plausible a conception of the ecosystem as a gigantic superorganism in its own right developing toward a mature, stable state of complex diversity (rather reminiscent of Clements' notion of a climax community).

So powerful was this image that, for a time, systems approaches seemed to be characteristic of ecology. That, at least, was the verdict of authors like Odum, writing in the late sixties. Moreover, the development of the mathematical theory of systems by Bertalanffy and others allowed this style of ecology to be compatible with the notion that what matters is mathematical rigour and precision, together with the development of mathematical models. It is now clear that at least some of the systems view is wrong, and that even where disproofs have not occurred in the field, the assumptions and methodology of systems ecologists need careful scrutiny. Part of the trouble is that we can find patterns wherever we look for them. With appropriate selection of habitats and species, it is all too easy to find a system of mature, diverse stability. At the same time, the biologist can only defend the conception of the ecosystem described above by not paying attention to massive counterexamples: for example, the lack of stability in tropical rainforests, and the many cases of apparently arrested, cyclic, or other *ad hoc* variants of succession where no climax system has emerged (Putman and Wratten 1984: Ch. 4).

Claims on behalf of a different style of ecological modelling have emerged in recent years. This consists in taking a more *individualistic* approach to communities. An individual plant, for example, will vary enormously in its attractiveness to a herbivore, and a tree will often show variation in its defences when attacked by invading micro-organisms. In this way, a tree or other plant, seems more like a population than an individual, for it constitutes a variable resource for herbivores and disease organisms. Over time, plants and animals show varying *life history* strategies. These again will vary, not just from species to species, but from individual to individual within a species. When this individualistic approach is coupled with appropriate statistical and stochastic models, the result is not necessarily an abandonment of the idea that there are ecosystems, or that competition and predation play their roles in shaping such systems. Rather, we have to recognise that there is a far greater complexity of levels at which description of natural phenomena can be attempted than was urged by theorists tied to one particular conception. Nor does the issue boil down to a dispute between atomistic and holistic approaches to the science of nature. Rather, as will be shown in the following chapters, there may be patterns in natural systems and plausible stories about the emergence and growth of such systems. But such stories need to be formulated with great care, and the patterns have to be shown to be in nature rather than foisted upon it by the eager interpreter.

Such a brief sketch of the history of scientific ecology is bound to be misleading. It can do little justice to the enormous richness of the subject, and has been silent upon major portions of the field's past. But I hope to have given at least a glimpse into some issues that have been of importance in the last hundred years or so, and to have provided a flavour of what the subject is about. In ecology, as in nearly all other cases, the best way to find out what it is, is by doing it. The sketch in this section has been no substitute for that.

4.2 Mathematical ecology

As already noted, there is nothing very new about mathematical approaches to ecology. With the development of systems theory and the increasing sophistication of mathematical modelling, what has happened in recent years is that some mathematical models have been developed which represent at least some aspects of real natural systems in a moderately plausible way. It is important to remember, however, that all modelling involves simplification which ignores particular complexities in search of general trends (for a useful survey of linear and non-linear models in ecology see Maynard Smith 1974).

Whole treatises could be written on the appropriate methodology for ecological studies. In particular, two issues arise for community and ecosystem study that are still very much in dispute. One is the question of the status and positing of null hypotheses (see Chapter 3.4). The other concerns the whole question of co-variance and correlation. If environmental factors really do affect the populations in a community — which simple common sense suggests they do — then a question arises over the appropriate statistical methods to be employed in charting the impact of biotic and abiotic factors on selected populations. How are we to collect and order the data on the species and populations in question? Which techniques yield the best fit among correlated data (this is a question about *multiple regression*)? The statistical questions raised here lie outside the scope of the present work, although reflections on the methodology of multi-variate analysis are aired in Chang and Gauch 1986.

In the rest of this section, I introduce a simplified account of some historically significant mathematical work in ecology. Further details on statistical models and kinds of explanation will be encountered in Chapter 7 — however neither that material, nor what follows gives a real indication of the technical difficulty of mathematical work in ecology. If you want to learn very much about mathematical ecology you will first have to learn mathematics!

Early mathematical models in ecology used deterministic rather than statistical or stochastic models. Deterministic models have the advantage of being easy to work with, although we have to bear in mind the simplifications they introduce. For instance, if an ordinary differential equation is used to describe predator–prey interactions, then even if that equation predicts a steady state of population density, any real system which is an 'instance' of the equation will show fluctuation round the steady state. Again a system for which equations predict large oscillations may see a population go extinct at a low point (Maynard Smith 1974: 16).

Early nineteenth century theorists were struck by facts about the *intrinsic rate of increase* in natural populations, and the limits imposed on this rate by the *carrying capacity* of the environment. Suppose that we measure the size of a population at regular intervals and find that, after starting at 20 it rises to 30, 45, 68, 102, 153, and so on. We give the initial population the value N_0, then subsequent ones are N_1, N_2, and so on, so that after t intervals our population is N_t. If we let $N_1 = N_0R$, we can call R the *fundamental net reproductive rate* of the population (Begon, Harper and Townsend 1986: Ch. 4.7). For our imagined population, the rate R is pretty constant, and stays close to 1.5. So long as R exceeds 1, population will obviously increase. We can thus derive the following equations for population growth over time in general:

$$N_{t+1} = N_tR$$

$$N_t = N_0R^t$$

From the second equation, we know that the population after three intervals will be the original population, times the rate of increase cubed.

Now suppose that we are interested not just in any time intervals but in those that are a generation apart. We will denote such

Figure 4.2: Exponential discrete growth

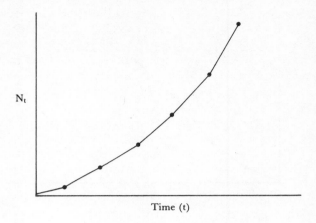

Time (t)

a number of intervals by T. So the calculation for obtaining the size of a population from its size one generation earlier is easy: $N_T = N_0 R_0$. Now, we can convert R, the fundamental net reproductive rate to a slightly different measure by defining $r = lnR_0/T$, where lnR_0 is the natural logarithm of R_0. Our new term r, the intrinsic rate of natural increase, is the rate at which the population increases in size.

Not all populations increase in discrete stages. For those that do, simple difference equations can be used to depict their growth, taking account of the limits of the environment. If we think about the equation $N_{t+1} = N_t R$, we will observe that this predicts indefinite and exponential growth as long as R has a value greater than 1. Thus, we could represent the growth curve for such a population as shown in Figure 4.2. Suppose, however, a point comes in the growth of the imagined population when competition for resources, attack by predator or other factors limit the growth curve so that the population at that point simply replaces itself each generation. The population size at which this steady replacement level is reached is one where N_t and N_{t+1} are equal. Such a population size can be defined as the *carrying capacity* of the system. The curve that takes account of carrying capacity (usually represented as K) is sigmoid rather than exponential (Fig. 4.3).

Figure 4.3: Discrete growth stabilising at carrying capacity K

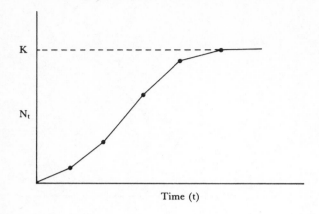

Time (t)

The equation for this curve is:

$$N_{t+1} = \frac{N_t R}{1 + \dfrac{(R - 1) N_t}{K}}$$

We are nearly in a position now to introduce the logistic and Lotka-Volterra equations, which lie at the heart of traditional work on population ecology. Let us complicate matters by thinking of continuous change through time instead of discrete jumps by breeding season. In order to model continuous growth, we need to use differential equations. This change in approach promises to make our models more realistic, since in many natural communities birth and death take place in a continuous fashion. Using the value *r*, as already defined (the intrinsic rate of natural increase), we can propose the following equation, originally suggested in 1838 by Verhulst, but often known as Pearl's *logistic equation*:

$$\frac{dN}{dt} = rN \frac{(K - N)}{K}$$

The continuous curve generated by this equation is shown in Figure 4.4.

58

Figure 4.4: Continuous growth limited at carrying capacity K

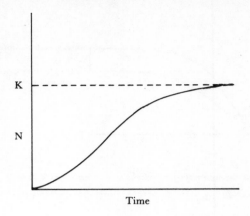

To think of a plausible interpretation, consider a hypothetical population in which competition for resources within it is slight while numbers are low but becomes ever more intense with increasing population density. At some point, then, growth in population hits a limit, for the rate of change of numbers with time (dN/dt) only increases while N is less than K. The rate will decrease for values of N greater than K and will be zero (that is, there will be no change) when $N = K$.

The elements of the Lotka-Volterra model are now to hand. Lotka, a student of Pearl, proposed in 1925 that taking the basic differential equation $dN/dt = rN$, we could build in features to account for competition within a species for limited resources, and also to account for competition among species. Volterra made use of a similar model one year later. We have already introduced the equation that deals with competition within a species (sometimes called the Lotka-Volterra 'prey equation'). It fits best a situation where intense competition for resources ('prey') leads to starvation and death within the population (Begon, Harper and Townsend 1986: Ch. 10.2.1).

For interspecific competition, let us imagine two populations whose instantaneous numbers are N_1 and N_2. The competitive effect of population$_2$ on population$_1$ is represented as $e_{1,2}$. We can now describe the growth of population$_1$ in the light of competition from population$_2$ by the following equation:

Figure 4.5a: Imagined effects of introducing interspecific competition

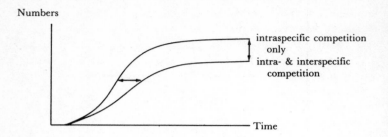

Figure 4.5b: How strong competition might lead to extinction

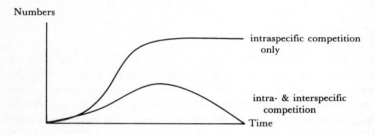

Source: Putman and Wratten 1984

$$\frac{dN_1}{dt} = r_1N_1\left(1 - \frac{N_1}{K_1}\right) - r_1N_1\left(e_{1,2}\frac{N_2}{K_1}\right)$$

The coefficient of competition, e, would have to be calculated for real populations in real communities — not an easy matter. But, sticking to the purely theoretical stance for moment, we can imagine different amounts of competitive effect. Consider the curves in Figure 4.5. As will be seen shortly, these theoretical possibilities can be made to fit simplified laboratory situations.

The introduction of the Lotka-Volterra model is not simply a means of studying populations. It also suggests ways of modelling communities and systems in general (see Chs. 8.3, 8.4). By starting from the simple analysis of intra- and interspecific competition, it is possible to refine the account to generate models that predict oscillation in population numbers according to resource utilisation. From this, we are led naturally to the study of

environmental variability, niche overlap, complexity due to
increasing the length of food web relations and the vexed issues of
equilibrium and stability in ecosystems and communities (for
example, Maynard Smith 1974: Chs. 5, 6).

To conclude the present brief treatment of mathematical ecology,
let us see how the Lotka-Volterra model copes with Gause's
Paramecium experiments. If we consider the results for the three
species of *Paramecium*, we find they can be arranged on the curves
shown in Figures 4.6 and 4.7. It seems clear that Gause's exclusion principle is no more than an informal statement of the Lotka-
Volterra model, as applied to a case like the one in Figure 4.7. An
interesting question arises for why *P. caudatum* does not eliminate
P. bursaria when they are grown together. The situation, of course,
is fitted by curves that can be drawn for inter- and intraspecific
competition; but the fit of the model *ex post facto* is hardly of great
interest. The biologist is liable to be intrigued by questions of why
the two species coexist at the levels found and whether altering
environing circumstances (for example, temperature) might make
a difference. Complicating factors involve changes during the life
histories of populations, time delays between resource consumption and appearance of new consumers, fluctuations in biotic and
abiotic environmental features, and so on. In other words, even
if we are happy to describe the laboratory data in terms of the

Figure 4.6: Growth of Paramecium *populations (*P. caudatum, P. aurelia*) in
separate culture (a) and in mixed culture (b)*

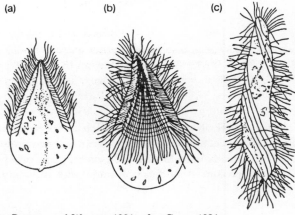

(a) (b) (c)

Source: Putman and Wratten 1984, after Gause 1934

61

Figure 4.7: Effect of competition on Paramecium *populations*

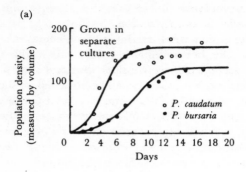

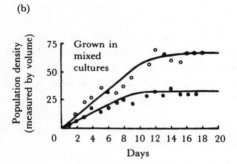

Note: Coexistence of *Paramecium caudatum* and *P. bursaria* in mixed cultures (b). Population curves for the two species in separate cultures are shown in (a). Source: Putman and Wratten, 1984

Lotka-Volterra model, it is unlikely that such a model can prove very explanatory in dealing with the complexities encountered in the field. As has already been hinted, the general question of what explains what in ecological studies is fraught with difficulty.

4.3 Connections

We are now in a position to consider whether scientific ecology can rightly be taken as the proper source of ecology as a metaphysic. It is already clear that some of the quotes from Gibson's and Capra's work in Chapter 3 echo features of scientific ecology. For example, there are indeed many levels of description at which we can approach the study of organisms and their relationships. We can think of organisms individually, as organised into communities with other organisms of the same species in a locality, and as organised into larger systems within which the populations of various species have well-defined roles. In all this, there is no attempt to find some fundamental unit in terms of which all the ecologists' descriptions can be given: in fact, it is unclear just what such a fundamental unit would be like.

As an almost immediate consequence of these observations, it seems that the findings of scientific ecology also support Gibson's general contention that ecological events differ from the events described by physical theory. In terms of overall conservation of mass and energy, very little may seem to be happening in a system of complex relationships among multiple plant and animal communities. However, we need to take care here. Gibson's original example concerned the change from ice to water — almost a non-event in terms of physical theory but a significant change from the standpoint of ecology. The point at issue may merely be one concerning the form of a description. In terms of physical theory, we may describe the movement of energy through an ecosystem in one set of relatively neutral terms. In terms of ecology, there will be issues of life and death, competition and resource partitioning. In its way, this vocabulary is also neutral (for example, 'competition' does not have the sense it has when applied to humans running races or throwing weights over bars). But there still remains a nagging doubt that these differences in description may amount to less than writers like Gibson think. The exploration of this issue leads us into the study of concepts of emergence and supervenience, and will be tackled in the next chapter.

There are, as we have seen, things that can claim to be distinctive ecological laws. Gause's principle of competitive exclusion, and the Lotka-Volterra equations apply to many different kinds of organism in many different circumstances. That they do so, may make it seem worthwhile to explore in more detail the concepts of competition, carrying capacity, niche, stability and so on which figure in them. At the same time, we have already seen that there are problems with such laws which do not arise for the principles of celestial mechanics or hydrodynamics. Because of the complex nature of ecology, it is hard to be sure that the proposed laws ever apply in an explanatory way to any real situation. The whole notion of what ecology explains, and what explanation involves, will occupy us in Chapter 7.

The further claims of Gibson and Capra are fraught with more difficulty than the ones so far mentioned. Is the duality of objective and subjective false? So far, nothing said about scientific ecology seems to suggest that the duality in question needs to be abandoned. It may be that closer inspection will reveal that what an item is, in some way depends on where it is located and to what it is related. As a special case of this phenomenon, the relation of observer and observed may depend on who is observing what and where. But the very practice of scientists is apparently at odds with the claimed relativity of objective and subjective. Gause did not regard the *Paramecium* competing with a rival protozoa as different from the one he grew on its own in the test tube. The difference in realised niche was not enough to lead to a change of species. Of course, it may be that some of my potential — given my theoretical or fundamental niche — is not being fulfilled in the niche I currently realise. Were I to be in some different situation, with different people and different resources available to me, perhaps I would be significantly different. In this way, perhaps humans differ from protozoa, or perhaps we just lack an appropriately detailed interest in protozoa. The influence of the community on the individual is a matter meriting further study, since its practical and moral dimensions are of some significance. So this is an area to which we return later.

Finally, two areas connected with Capra's metaphysics are so far untouched by scientific ecology as described here. One of these will be explored in the next chapter. The other will be left almost entirely aside. The first issue concerns reductionism. Capra's anti-reductionist claims need clarification, but in one obvious sense the thrust of scientific ecology is reductionist. In developing concepts that can apply to different kinds of community and even to individual organisms — like the concepts of competition and of niche — scientific ecology allows that laws and principles applying in one area will also apply to another. In a sense, behavioural patterns among hummingbirds 'reduce' to those found among protozoa. We will return to this later.

There will be no return to the claim, however, that patterns of nature are merely reflections of the patterns of the observer's mind. Such a thesis needs more than scientific ecology to inspire or support it. There are, of course, perfectly good philosophical supports for such a view. Kant, for example, argued that the mind organises experience in a dual way. Our sensory input, on his view, is structured by certain forms of sense, and our conceptual capacities are structured by certain categories of the understanding. Without some such philosophical support, there seems little merit in arguing for Capra's idealism. It will therefore not figure in my characterisation of metaphysical ecology.

Later in the book, I will be considering other views that draw some support from scientific ecology. Lovelock's *Gaia* hypothesis, for example, is a kind of rebirth of Clements' superorganism idea, extended to the biosphere as a whole. There is an interesting amount of confirmation available for Lovelock's claim, but our assessment of it must wait until we see the relevance and use of such confirming evidence to particular hypotheses.

A special kind of environmental egalitarianism also draws support from scientific ecology. This position, often known as 'deep ecology', provides a comprehensive framework for environmental decision-making, together with recommendations about good forms of life. The deep ecology programme is closely coupled with a perception of where human beings are located in nature and in natural systems. Deep ecologists tend to be holistic and anti-reductionist, and so it is prudent to consider their claims only after getting clear on these latter issues. So I now turn to a specific study of the controversy surrounding holism and reductionism.

5

Reduction and Holism

5.1 Parts and wholes

One natural way of understanding how a complex object or system works is by taking it apart. With the pieces all displayed, scrutinised and further dismantled, we are — with luck — in a position to venture suggestions about how they interact to bring out the characteristic behaviour and properties of the whole of which they were originally parts.

It has often been said that science is a rigorous and systematic extension of common sense. So it is hardly surprising that much scientific investigation of natural objects and natural systems involves dissection, separate investigations of separable components and hypotheses about the way such components combine and interact to yield the features of the object or system under scrutiny. However, associated with this analytic approach, some would claim, has been a neglect of features of whole systems, and an attempt to reduce novel properties and characteristics to interactions and behaviour involving the already familiar.

These concerns about the nature of scientific explanation and about the existence of various levels of phenomena receive a natural expression in several studies of biology. The degree of concern, however, varies from author to author. Thus Skolimowski in his book *Eco-Philosophy* simply objects to what he regards as the undesirable compartmentalising of our knowledge. He writes:

> The separation of facts from values, of man from his knowledge, of physical phenomena from all 'other' phenomena, resulted in the atomization of the physical

world, as well as of the human world. The process of isolation, abstraction and estrangement [of one phenomenon from other phenomena], a precondition of the successful practice of modern natural sciences, was in fact a process of conceptual alienation . . . The primary cause of contemporary alienation is a mistaken conception of the universe in which everything is separated and divided. (Skolimowski 1981: 14)

These claims seem modest by contrast with some of the remarks of others who criticise the atomistic nature of modern science. Thus Capra writes:

Quantum theory has shown that subatomic particles are not isolated grains of matter but are probability patterns, interconnections in an inseparable cosmic web that includes the human observer and her consciousness. Relativity theory has made the cosmic web come alive, so to speak, by revealing its intrinsically dynamic character, by showing that its activity is the very essence of its being. (Capra 1983: 83)

Whereas Skolimowski is giving a diagnosis for a kind of alienation he thinks exists between the knower and what is known, Capra is apparently concerned with the nature of the elementary particles themselves. They are not isolated 'atoms', but are interconnections in a web or field.

In the second chapter of the book from which the remarks above are quoted, Capra argues further that there is a connection between thinking atomistically and adopting the kind of analytical approach he discerns in the work of Descartes. As we saw in the first chapter, Gibson's theory of affordances likewise challenges an atomistic account of visual perception. The affordance of the environment — what it offers the organism for good or ill — are in Gibson's view equally part of the environment and a fact of behaviour (Gibson 1979: 127–9).

The views mentioned so far are not aberrations. Others have made more extreme claims and the more we examine these, the greater becomes the need to sort out what such writers are trying to say. As we will see in the course of the present chapter, there are a number of very different things that might be said about wholes and parts, about the nature of explanation and about the existence of laws and properties at various levels. For a final example, however, let us look at the following remarks by Lovelock. They are made immediately after he introduces the electric oven as a down-to-earth example of a *cybernetic* or self-regulating system:

> Think again about our temperature-controlled oven. Is it the supply of power that keeps it at the right temperature? Is it the thermostat or the switch that the thermostat controls? Or is it the goal we established when we turned the dial to the required cooking temperature? Even with this very primitive control system, little or no insight into its mode of action or performance can come from analysis, by separating its component parts and considering each in turn, which is the essence of thinking logically in terms of cause and effect. The key to understanding cybernetic systems is that, like life itself, they are always more than the mere assembly of constituent parts. (Lovelock 1979: 52)

There is a lot wrong with what Lovelock says here. We can recognise some of the peculiarities in what he says by reflecting on the fact that it is precisely an understanding of the causal role of each of the oven's components that would enable us to diagnose what has gone wrong in the event of a breakdown. Indeed, the parts of a complex artefact are located and designed precisely so as to enter into appropriate causal relations with other parts. But it is important to try to understand what Lovelock and others are trying to say when they maintain that the whole is more than the sum of its parts.

In the immediately following sections, I will not be carrying out any novel philosophical explorations. In fact, the issues being confronted in this chapter are well understood, although they bear on matters of extreme difficulty. So what I will do first of all is supply some clear examples of how we can distinguish different kinds of reduction, and try to show ways in which it makes sense to speak of wholes as distinct from, and more than, their parts.

5.2 Varieties of reduction

'Reduction' and 'reductionism' are terms understood in most philosophical treatments to apply to terminological matters. In the history of logic, for example, one of the most famous examples of reduction was the discovery that the logic of sentences (that is the logic involved in inferences whose validity depends on terms like 'and', 'or', 'if' and 'it is not the case that') could be carried out using only one primitive sentential connective. We can get some of the flavour of this discovery if we think about the relation between negation, conjunction and disjunction.

Think of the conditions under which the sentence 'Patricia is tall and Quentin is short' is true. Clearly, it is only true provided both its component sentences 'Patricia is tall', and 'Quentin is short', are true. Indeed, if either of these component sentences on its own is false, then the whole sentence built up out of them using the connective 'and' will also be false. What we have just said about the conditions under which the whole sentence is false can be put as follows. The whole sentence is false if either it is not the case both that Patricia is tall or not that Quentin is short. So now think of the following sentence: 'It is not the case both that Patricia is not tall or that Quentin is not short'. Although it is somewhat cumbersome, this latest sentence expresses precisely the same claim as our earlier one.

What we have just shown is that the term 'and' can be replaced in certain contexts by use of the terms 'or' and 'it is not the case that'. Put in symbolic terms, using 'P' and 'Q' to represent our two sentences, and using the signs '&' for 'and', 'v' for 'or' and ' ⅂ ' for 'it is not the case that', we can demonstrate the equivalence of any conjunction with a sentence constructed using only negation and disjunction and thus establish this equivalence:

$$P \& Q = \text{df} \ ⅂(⅂P \, v \, ⅂Q)$$

In the light of this definition we have *reduced* the 'and' of conjunction to the connectives of negation and disjunction.

Since the logic of sentences is usually carried out using five connectives, our result here shows how to reduce three of them to a mere couple. The general result, discovered at the turn of the present century, is that all sentential logic can be done by one of two special connectives. Provided we are prepared to write sentences in rather long, and repetitive forms we can dispense with symbols for 'and', 'or', 'not', 'if' and 'if and only if'. The connectives that enable this wonderful reduction to take place are joint denial ('it is not the case that both P and Q') and disjunctive denial ('either it is not the case that P or it is not the case that Q'). As the reader can easily ascertain, we pay a high cost for such reduction in terms of comprehensibility and length. But such considerations do not affect the principle of the procedure.

The case given is one of the simplest examples of reduction. In more complex cases, we can eliminate certain symbols without even giving explicit definitions of the terms eliminated. Thus in Russell's famous treatment of definite descriptions, it was proposed that certain kinds of descriptive phrases could be eliminated from sentences, even though the procedure for bringing about this elimination did not involve the replacement of the descriptive phrases by any others doing the same work. In our earlier example, the work of 'and' is done by a combination of 'not' and 'or'. The symbol '&' is thus definable in terms of the symbols ' ⊣ ' and 'v'. But in Russell's reconstrual of sentences like 'The present King of France is bald', no symbol or set of symbols does the work of the phrase 'the present King of France'. Rather, his theory of descriptions involved replacing the whole sentence with another sentence which does the work of the original one: so his theory gives us only a *contextual* definition of descriptive phrases.

These classic examples of symbolic reduction, based on definition, reveal the core of what philosophers of science usually mean when talking about *reduction*. It should be noted that reduction thus understood is not simply a matter of noting a correlation between truth conditions of two classes of sentences or statements. Russell's theory, like the reduction of the five logical operators, tells us at least something about the *meaning* of the reduced sentences (as is argued for reduction in general by Dummett 1978). But when we turn to statements in a scientific theory and consider the prospects for symbolic reduction, we find a much more complicated situation. We can get a grasp of some of the problems by thinking about a simple case from mathematics — the reduction of theorems and terms of arithmetic to those of set theory. Although this is sometimes described as the reduction of arithmetic to set theory, it is still a case of reduction of symbols.

Arithmetic, based on the relationships among natural numbers (0, 1, 2, 3, . . . and so on) has a number of laws which seem, on the face of it, similar to laws in other areas. For example, the associative and commutative laws for addition have parallels in set theory. The suggestion, pursued by von Neumann, was that somehow all the laws of arithmetic could be expressed as laws of set theory: arithmetic could then be deduced simply from set theory. What was needed, to deliver this result, was some way of translating terminology making reference to natural numbers into terminology making reference to sets (just as we have shown ways above of translating the terminology of conjunction into the terminology of 'or' and 'not'). The way von Neumann did this was extremely simple, although highly ingenious. He took that most important number, zero, to be defined as the empty set. The empty set, like 0, is unique: it is the only set which contains no elements. With this definition of zero to hand, the rest of the reduction was straightforward. Any natural number was defined simply as the set whose elements were all preceding sets. Thus one was the set whose only element was the empty set, and two was the set whose only elements were one and the empty set.

With such definitions available for each number, von Neumann was able to recast all the axioms and theorems of arithmetic as axioms and theorems of set theory. Notice that in thus reducing arithmetic to set theory, von Neumann had taken a step that was not present in our earlier examples. In reducing our logical vocabulary to smaller and smaller numbers of connectives, we were working all the time within logic. There are logical techniques for showing that 'not' and 'or' can do the equivalent work of 'and'. Moreover, even in the Russell case, there are techniques of a sort that can show that, at least on one understanding of sentences containing descriptions, the sentences his theory delivers are indeed equivalent to the original sentences. Within these special, logical systems, then, there exist techniques for showing that the reductions are meaning-preserving (and thus worthwhile if we are interested in semantic or conceptual economy). But in the case of arithmetic and set theory, there are no set theoretical or arithmetical techniques that will reveal the truth of the vital claims that zero is the empty set, that one is the number whose sole member is the empty set, and so on. These are, in fact *postulates* made with an eye to securing the required reduction and the consequent derivations of arithmetical laws within the set theory. If we dissent from the postulates, or if further investigation revealed that they are not plausible, then we would have grounds for dissenting from the proposed reduction.

There is, however, one consideration that makes the reduction of arithmetic to set theory less problematic than reduction in the empirical sciences, namely that we have no very clear grasp of what numbers are in the first place. Thus there are alternative ways of defining numbers in terms of sets, which all deliver the desired reduction of number theory to some version or other of set theory. The attempt to find some means of reducing all of higher mathematics to set theory, and then ultimately to logic, is at the heart of the programme of *logicism*, investigated by both Frege and Russell. There are technical reasons for doubting that the logicist programme could ever come off. But if it were possible to effect the desired reduction in one way, it would not be surprising if it were also possible to effect it in some other way as well. Our unclarity about what numbers and other mathematical objects are would make us amenable to alternative reductions of mathematical vocabulary.

In the paragraph above, I have at last moved from considerations about mere terminology or symbols, to consideration about things referred to by the symbols in question. For to make sense of the claims just made, we have to agree to something like the following points. First, there are terms or symbols used in mathematics which refer to certain concepts — numbers, functions, vectors, and other mathematical entities. Second, since we are unclear about the nature of these concepts, we will be willing to accept alternative accounts of how sentences expressed in mathematical terminology are to be expressed in purely set-theoretical or logical terminology. The rationale for the point about vocabulary, then, involves making explicit reference to the items we are using that vocabulary to describe.

This is important. What has just been argued makes it look as if there is the most intimate connection between claims about terminology on the one hand, and claims about what exists — ontological claims — on the other. It looks as if in reducing the *vocabulary* of arithmetic to the vocabulary of set theory we are doing away with arithmetical *entities*, and showing that numbers do not really exist at all; the only things that need to exist for the laws of arithmetic to be true are sets, and sets of sets. There are thus two kinds of reduction to be distinguished although it looks as if they are related. First, there is the reduction of terminology or symbols; second, the reduction or removal of entities. Let us follow normal practice, and reserve the term 'reduction' for the former kind of occurrence. And let us call the removal of entities 'ontological reduction'. What I want to show in the next section is that, contrary to first impression, the two kinds of reduction are not connected at all. In fact, they are — surprisingly — independent of each other.

The independence is particularly clear in the cases of reduction which involve not a *translation* from one class of statements to another but are concerned merely with truth. In this weaker sort of reduction (the 'reductive thesis' of Dummett 1978) statements in the one class are only true if statements in some other class are also true. The truth of certain statements in the reductive class then guarantees the truth of corresponding statements in the correlated class. This type of reduction can still be of semantic significance, and thus be of importance to our understanding of certain kinds of sentence. But once we see that the connections involved between the two classes of statement are simply ones involving truth conditions, we are less inclined to think of the reduction as having any ontological significance.

5.3 Occam's razor

What I will now show, by simple examples, is that the reduction (in either the weak or strong senses already identified) of one theory's vocabulary to the vocabulary of another theory shows us nothing about the number of things, or kinds of thing, there are in the universe. Then I will also show that, in a universe in which there is only one sort of thing — namely, physical things — it can still turn out that a complete description of some part of it will have to use predicates that are not physical predicates by any stretch of the imagination. After that, I will cast doubt on some well known claims about levels or hierarchies of things.

So let us think first of all of the point about reduction in terminology. It was argued by Descartes, Leibniz and the other rationalist philosophers that minds were an entirely different kind of thing from bodies. Their picture of what exists in the universe, their ontology, thus allows for the existence of two quite separate categories of thing: no body is a mind, and no mind is a body; the properties of mind include thinking, desiring, willing, and so on, while the properties of the body involve physical extension and causal interaction with other bodies. Now, a natural way to describe the dualist's universe is by distinguishing those things that are mental from those that are physical. Indeed, it might seem essential to stating the dualist's view of the world that we are able to use the terms 'physical' and 'mental'. But this is not so at all. We can give a description of the dualist's universe while avoiding the term 'mental' altogether: for we simply replace all occurrences of this term in our descriptions by the term 'non-physical'. So we here have achieved reduction in vocabulary: but nothing at all follows about removal of entities. We have not succeeded, by this move, in reducing the ontology of the dualist to one where there are only physical things.

Of course, real reductions in science are rather more complicated affairs than the trick given here. But the simple point is compelling. Let us look, however, at a well known case of reduction in science to see if there is room for claiming that entities have not been removed despite a real reduction in vocabulary. It is well known that some branches of science have a comprehensiveness about them that makes them applicable to a diverse range of subject matters. A classic example of the comprehensiveness of statistical mechanics is its ability to reduce thermodynamics. Nagel gives a careful description of one part of this reduction in his book *The Structure of Science* (Nagel 1961: Ch. 11). The kind of terms that characterise the laws of thermodynamics — 'heat', 'entropy', 'temperature' and the like — are replaced under the reduction by terms characterising the mechanical behaviour of molecules and aggregates of molecules. But, of course, we cannot give any account of laws involving pressure and temperature in terms of mechanics until we are able to specify by virtue of what features gases show mechanical behaviour (for example, that they consist of vibrating molecules, whose impacts on the wall of any container constitute the pressure of the gas), and how the temperature of a gas is associated with energy of its constituent molecules.

So what happens in the reduction of thermodynamics to statistical mechanics is that a number of auxiliary assumptions or postulates are introduced to link the two areas, and give appropriate content to the derivation. As Nagel puts it (dealing with the reduction of the law that pressure of a gas is a function of volume and temperature):

> . . . the Boyle-Charles' law is a logical consequence of the principles of mechanics, when these are supplemented by a hypothesis about the molecular constitution of a gas, a statistical assumption concerning the motions of the molecules, and a postulate connecting the (experimental) notion of temperature with the mean kinetic energy of the molecules. (Nagel 1961: 345)

We have here a case rather like the one involving arithmetic and set theory, but also significantly different. In that case, there were postulates that gave content to the notion that numbers could be regarded as sets. Likewise, in the present case, there are postulates that explain how temperature, for example, is to be taken as mean kinetic energy of molecules. However, we no longer have either a straightforward *translation* of one vocabulary into another, nor even a semantically significant *correlation* of truth conditions between statements using the two vocabularies.

As Nagel and other writers have pointed out, the additional postulates are by no means statements of pre-existing synonymy, or even partial specification of the meaning of a term like 'temperature' (see the discussion in Hempel 1966, quoted in Chapter 8). There are, certainly, other terms that are common to both mechanics and to thermodynamics (for example, 'work', and maybe even 'pressure'). But in interesting scientific reductions, not all the terms belonging to the reduced and the reducing theories can be common (or else there will be no possibility of reduction in the first place). Conversely, there can be no valid derivation of the laws of one discipline from those of another discipline where the derivation introduces new terms into the conclusion that were not already in the premises. (We ignore such cases as the derivation 'All metals expand when heated; therefore all metals expand when heated or hummingbirds sip nectar'.) It follows immediately that extra assumptions, or principles of connection, are required in any interesting reduction, to relate the distinctive terminology of the two sciences.

That these extra assumptions required in interesting cases of reduction are substantive and significant does not mean that they are factual. It may be that we encounter independent evidence to show that temperature is associated with the mean kinetic energy of gas molecules. By contrast, we might never make any such empirical finding, but decide that taking temperature to vary with mean kinetic energy of molecules is justified by other features of the reduction — for instance, we find empirical evidence for other additional assumptions, and the reduction exploits existing parallels between the two areas.

It might be thought that we can argue that at least some kind of weak reduction occurs in the cases currently under discussion. But even the 'reductive thesis' requires there to be some semantic connection between the reduced and the reducing statements. However, it need not be maintained that our grasp of sentences about temperature depends on even an implicit grasp of sentences about kinetic energy. Indeed, it would be perverse to maintain any such thing. So we might argue that in such cases we have reduction in some still weaker sense, marking it by using scare quotes. What we can do is 'reduce' thermodynamics to statistical mechanics by 'translating' sentences about temperature into sentences about mean kinetic energy. Of course, if the reduction becomes so well established that our preferred way of thinking about temperature and pressure is in terms of kinetic energy, then the situation starts to change. We can then so arrange matters that a learner's grasp of temperature and pressure phenomena involves a grasp of mechanics so that, at least within this specialised sub-framework, there come to be semantic connections between statements in thermodynamics and statements in statistical mechanics. This would still not establish that there is any general semantic connection between everyday talk about temperature and the theory of mechanics.

With this case still before us, we can now ask if anything of ontological significance has been achieved in the reduction. The answer plainly is negative. What the reduction shows is that the pressure of a gas can be thought of as a function of mean kinetic energy of its molecules and temperature; moreover, the equation for this function can be deduced within mechanics. And the finding shows us something about the temperature of gases — provided we accept the postulated relation between temperature and mean kinetic energy. Of course, such a reduction might encourage us to go out and look for independent evidence concerning the postulate. But, lacking it, we could still regard our understanding of temperature and pressure phenomena enriched, given our prior understanding of mechanical phenomena.

The case thus bears on questions of explanation, understanding and the motive for further experimentation. But it has no obvious bearing on ontology. The temperature of a gas is one of its properties or characteristics. It is measurable in a number of ways. But it is not something existing alongside the gas. In this way, the case differs interestingly from our case involving arithmetic and set theory, for it looked in that case as if we had grounds for removing numbers from our ontology and thinking of the objects of arithmetic as being sets, and sets of sets. It may seem that there is the possibility of some ontological reduction here, however. For does the reduction of laws concerning gases to laws concerning aggregates of particles not suggest that gases are no more than aggregates of particles? This amounts to the proposal that gases be regarded as 'ontological parasites' in Chisholm's phrase. The issue of ontological parasites is a confused and confusing one, and I deal with it at the start of the following chapter.

Let us conclude this one by looking at the second claim I set out to establish: that in a universe of only one kind of thing, a complete description may require terminology that apparently characterises different kinds of thing. This can be shown quite easily if we follow Boyd's account of the plasticity of the mental. Suppose, then, we agree, at least for the purposes of my argument here, that the same mental state can have different implementations in different physical systems. Mental states are thus like computational states, displaying what Boyd calls 'compositional plasticity' (Boyd 1980). Thus, although — on a materialist account — having a pain will always involve a state of a body, a qualitatively similar pain might be constituted by quite remarkably different physical and chemical arrangements. The same goes for states like memories which, let us suppose, are records that an organism keeps of its past experiences. Now, it may be that in order for it to have certain experiences at all, there have to be some physical things in an organism's environment. Nonetheless, there may be quite distinct ways in which an organism might keep track of its environment, involving a causal relationship with objects of previous experience which are not at all like the ways our best theories of memory think this record-keeping is in fact done.

An extreme consequence of this way of thinking is that we might be materialists while conceding that it is possible for there to be mental states that are not given any physical implementation at all. Materialism, as understood here, is no more than the claim that certain physical structures and processes happen to constitute mental states as we know them. Such a modest materialism thus denies that there will be any formula or set of formulae for translating mental state predicates into physical state predicates (like the traditional case of 'John has a pain' being translated as 'John's C fibres are firing'). On the other hand, this version of materialism does insist that, given the way the world actually is, some physical states will constitute states of remembering, intending, feeling pain and so on. Thus the world as it actually is contains only physical things: yet some of these are also mental things (Davidson 1982: Essay 11).

So long as we are able to make sense of this modest version of materialism, we are able to note two important facts. One is that, as we have seen, successful reduction need not have any ontological consequences. The second point, just stated, is that in even a purely physical world it may be impossible to reduce our terminology to merely physical terminology. Together, these points have a devastating effect on complaints about reductionism. For they allow us to keep questions of ontology quite separate from questions about terminology and meaning. In cases where the two domains seem connected — as in our example of the reduction of arithmetic to set theory — there had, therefore, to be other factors involved. So indeed, it turns out, is the case. A number of reasons make the reduction of numbers to sets seem attractive, not least the methodological principle of Occam's razor: entities are not to be multiplied beyond necessity.

As is obvious in the case of the reduction of arithmetic to set theory, and as has been noticed by other writers (for example, Quine 1966, Ch. 17; Wright 1983: Ch. 1), there is no *a priori* reason why the reduction should not work the other way round. Think of it like this. The von Neumann reduction allows us to specify a function (what Quine calls a *proxy function* in Quine 1966: 205). This proxy function enables us to specify, for each of the things of which a number-theoretic predicate is true, the corresponding item of which a set-theoretic predicate is true. Further, for the primitive predicates of number theory, the function assigns to each sentence satisfied by one or more numbers a corresponding sentence satisfied by one or more sets.

But now think of the whole process working the other way. Suppose we are able to devise a proxy function that maps sets and their predicates onto numbers in just the way described above. What we would have done by this is reduce set theory to number theory. So we need more than the existence of a technique for reduction to convince us that any particular reduction should be taken as ontologically significant. For all we know there may be numbers and there may also be sets. Or there may simply be numbers, but no sets. Or there may be sets, but no numbers. These are all open ontological possibilities for a philosophy of mathematics. What Occam's razor suggests, then, is that we would need some special reason for claiming that there are both sets *and* numbers. What it does not do is sort out whether we should make our ontological reductions in one direction rather than another.

Of course, there is general consent among philosophers of mathematics that reduction in one of these directions is sensible and worthwhile. The comprehensiveness of set theory, its ability to reduce other areas of mathematics, and the prior existence in it of results that are isomorphic with results elsewhere encourage reduction to sets. Likewise, we might similarly argue that the comprehensiveness of mechanics, the existence of laws in mechanics that are already pretty similar to laws in other fields and its ability to regiment discussion in other fields make reduction to mechanics a plausible option. Of course, the reduction of temperature to molecular kinetic energy is not this time a reduction of entities in one domain to entities in another. As we have already observed, there was no ontological significance in that particular reduction.

It should by now be clear that the issue of reduction is an extremely complicated and difficult one. Not all reduction involves ontology. Nor is there any *a priori* reason for taking terminological reduction as having ontological significance. Moreover, we have found that in one case where there was ontologically significant reduction, the entities in question were both of a somewhat peculiar kind: they were both mathematical entities, entities of theory. In the next chapter, more will be said about the status of theoretical entities.

In the meantime, though, let us consider a phoney argument for reduction. On the face of it, it look appealing, and in considering it we come closer to issues raised by the writers quoted earlier in the chapter. For it should be clear that the technical issue of reduction does not offer any easy key to the solution of the question of wholes and parts. However, inspired by the example of the inclusion of thermodynamics within mechanics, someone might venture the following argument. If the temperature of a gas is simply constituted by the vibration of gas molecules, then why think of gases as being real at all? Why not recognise that gases are mere collections of vibrating molecules, and recognise that all our talk about properties of gases is therefore talk about such collections. This strategy introduces a new element into the discussion. For it is an attempt to persuade us that some items are, in Chisholm's phrase, 'ontological parasites' with regard to certain other, more fundamental, items. In addressing this point, and displaying the confusions it involves, we will start to become clearer on the part–whole issue. Note, however, that in discussing ontological parasitism, we are not discussing reduction in any of the forms so far identified.

6

Nature and Existence

6.1 Ontological parasites

It is tempting to be lured into a significant mistake by the following chain of reflections. My desk, upon which the machine I am using now stands, is a collection of parts — a top covered in vinyl that is meant to look like leather, four legs, some pieces of beech trim, wooden blocks, screws and various portions of set glue. These parts, or components, are the sort I would come upon separately were I to dismantle the desk in a careful and systematic manner. But a little reflection will convince us that each of these parts itself consists of yet other things. The legs, being of solid beech are made up of wood fibres, themselves consisting of the long cells typical of plant material. These cells, however, are not unstructured, but also have their own components. Ultimately, we might think, the desk is a swarm of molecules, with large spaces between them, their aggregate vibration being what supports the word processor at which I now work.

These reflections have led us from the familiar desk of common sense to Eddington's celebrated desk of physics. To see what is wrong with them, however, we do not need to depart beyond the first step in the journey just described. Let us suppose, then, that it is suggested that the desk is simply a collection of desk parts. Thus what is supporting the keyboard before me is a collection of desk parts. The desk itself is an *ontological parasite* on its parts (Chisholm 1976: Ch. 3). Put the point another way. Desks are not items that exist alongside, so to speak, collections of desk parts. Rather, to think about desks is a convenience. We have the concept of *desk* because to work with such a concept is more convenient than working in terms of the concept *collection of desk parts*.

Nonetheless, in a complete inventory of the universe, desks would not have a place, so long as we list all the desk parts.

There is something right about this suggestion, but also something misleading. What is right is the thought that desks are constituted by certain parts. But it would be wrong to think that a desk is *merely* a collection of parts of a certain sort. In fact, there are many collections of desk parts that are not desks (for example, the collections to be found in factories where desks are put together). Many of these collections will never be desks, even though some of them might have been. Someone who defends the ontological parasite view might agree that not every old collection of desk parts is a desk, but argue that there are not desks *in addition to* desk parts. Let us suppose that there are 18 items in our preferred list of desk parts. The ontological parasitism of desks might be revealed by pointing out that there are not 19 items involved in supporting my keyboard — namely the 18 desk parts and, additionally, the desk.

To see what is odd about this way of thinking, consider the way that someone might argue for the claim that desks are distinct from their parts, while denying that 19 things now support my keyboard. To be a desk is not simply to be a pile or collection of parts. The parts have to be causally integrated in an appropriate way, involving screws, nails and glue. The legs of my desk support the top, while the top also separates the legs — these are further causal relations the components have to each other. Thus a desk is a causally integrated, structured collection of parts; it is a unitary object (so that if you interfere in certain ways with one part, your interference affects the others); it is movable separately from the things around it and it has predicates true of it that are not true of any of the parts taken separately.

The point is that in rearranging, connecting and modifying the materials around us in the world, we make new things. This point we can maintain even while allowing that the things thus constituted are made solely of ingredients that were already there. But it would be absurd to maintain that what is really there is simply that material upon which we work to produce artefacts, and that artefacts are not themselves anything but parasites on this material. For the claim that artefacts like my desk are *real* is not the claim that in manufacturing such things we add to the basic stock of *materials* in the world. The claim that artefacts are real is the claim, to put it simply, that they are subjects of predication or, in a more material mode, the claim that they have properties in their own right. Of course, the theorist impressed by ontological parasitism can now maintain that what supports my keyboard, and may be moved to my new office, is a unitary, causally integrated and structured collection of desk parts. But this is simply to grant my point, for there is nothing parasitic about such structured collections: they are precisely what desks are.

The truth in the claim that tables are not additional to table parts is the truth in the claim that structure, and causal relations of the sort that give an item its unity and integration are not additional elements, or entities, taking up space in the world alongside the parts of things. But nothing follows from this claim about the reality, non-reality or ontological superfluousness of whole objects. At this point, the defender of the view that parts are more basic than the wholes they constitute may make one of several moves. One is to turn again to the possibility of reduction. It might be said that a description of the world that does not mention tables is possible: for every claim about a table can be reconstrued as a claim about a structured, integrated collection of parts. But, as shown in the last section, nothing follows as such from this sort of terminological point. That talk about tables can be reduced to talk about collections of parts in no way establishes that there are no tables.

Incidentally, it is not clear that even the reduction of table talk should be allowed to the imagined opponent. For not all things that count as tables have clearly defined top, legs, connecting pieces and so on. Indeed, since for something to be a table is for it to afford us certain uses and to function in certain ways, there is a certain plasticity about tables, just as there is about mental states. Tables, to put the point pretentiously, can be implemented in very different physical systems. So it is not clear that any term formulated to apply to integrated collections of parts would apply to all and only those things that are tables.

My conclusion, then, is that in order to be a physicalist, one does not need to hold that certain categories of physical things are more fundamental than others. It may be that there are only physical facts, and that the physical facts exhaust and determine all the facts there are (Hellman and Thompson 1975). But such considerations in no way undermine the reality, or substantiality of items that are complex. Indeed, any complex item must be a collection of parts: if that item merits inclusion in some kind or sort that we recognise, then it will have the structure and causal integration appropriate to the kind to which it belongs.

What sense, then, can we give to the contention that the whole is more than the sum of its parts? There are a number of boring points we can make here. First, unless the word 'sum' is being used in a special sense, a whole is never merely a sum of parts, for it needs to be integrated and structured in the ways already mentioned. For this reason, we distinguish things that are of genuine sorts or kinds from those that are not. My table is something of an artefactual sort, and the tree beside the compost heap is something of a natural sort or kind. But the sum of the tree and the compost heap is not a unified item of any sort at all — although it undoubtedly is a sum of parts. Likewise, my desk and the door of my office together constitute a sum of parts, but are not any one thing of a kind.

Those philosophers committed to a policy of extensionalism will want more by way of evidence for these last claims (which can be found in Ch. 7 of Brennan 1988). But the rest of us are perfectly clear not only that trees are genuinely of certain sorts or kinds, but also that the property of being a tree is a genuine property in the way that the property of being, at one and the same time, a tree or a compost heap is not a genuine property at all. One of the important problems facing the philosophy of ecology, as we will see, is the question of whether ecosystems are items that are more like trees, or items that are more like a sum of fairly unrelated parts. This topic will be discussed in the next chapter when we look at the status of theoretical entities.

One last move that might be made by the person wedded to the theory of ontological parasites would be to argue that wholes depend on parts in the following way. If the parts did not exist, then the whole would not exist either. Such a dependency relation threatens to make the wholes seem less real than the constituent parts. Indeed, the move is familiar, in another guise, in the history of philosophy. Descartes, for example, argued that what he called *substances* (things) were more real than *attributes* (general properties) or *modes* (more specific properties). Moreover infinite substance was more real than finite substance. Again, there was a supposed dependency. He thought that in order for a property to attach to something, there had to be a thing for that property to attach to. Things (substances) are thus more real than their properties. Indeed, the properties of something can change while yet the thing stays the same. Descartes' notion of a hierarchy of reality is not unlike the imagined hierarchy of entities suggested by the parasite theorist. The development of Descartes' point simply takes us in a different direction.

The simplest reply to this latest argument is that it is seriously question-begging. Let us think of one of the legs of my table. It is a piece of beech, roughly 2 × 1 inches in section and 27 inches long. Take it away from the table and then ask yourself the question: what is it? Is it a shelf support, a bracing strut, a portion of a picture frame, a piece of window sill, part of the frame of a settee, or just a piece of timber? There are clearly a number of functions it might perform, but no functional description rather than another is appropriate to it in the absence of further information about its location in a structure involving other components and — perhaps — without information about the design intentions lying behind its manufacture.

The table leg is a relatively uncomplicated thing. When we turn to more complex components — whether nested in artefacts or in natural objects — we find that some are so function-specific that, given enough knowledge, we can tell from inspection of the component just where it comes from and what role it does. An exhaust manifold or a twin choke downdraught carburettor are, to the knowing eye, obviously components from internal combustion engines, just as acorns are the fruit of oaks. It would be odd to argue that acorns and carburettors are more fundamental than the things of which they are parts. For without internal combustion engines and oak trees these things would not exist either.

The difference between unified wholes and mere sums of parts is not directly linked to the question of reality. For what I have been arguing is that wholes are just as real as their parts — no more so and no less so. Thus the whole consisting of the tree together with the compost heap is no more (and no less) real than the tree, the compost heap or their respective parts. This is just what we would expect if we abandon the doctrine of the ontological priority of parts. One difference between unified and non-unified part collections is that only the former are relevant to the functional status of components. The acorn is the fruit of the oak tree, not of the tree and the compost heap. The carburettor has a role in the engine, not in the engine taken together with a telephone, even though the engine and telephone do constitute a sum of parts.

6.2 The properties of wholes

If structured, causally integrated, unified things are more than a sum of parts, then this will be revealed in a number of ways. First of all, think of ways in which this will *not* be revealed. A sum of parts will take up a determinate volume of space, as will a whole object — indeed, the same volume. So the difference between wholes and sums of parts cannot reside simply in terms of their occupancy of space and time. In fact, it is going to be very hard to say just what the features of unified things of a kind are: for these are connected with their having genuine properties.

The influence of formal logic on philosophy has not been very helpful in this area, for our grasp of what a property is cannot be elucidated by logicians' accounts of predication. Let us, for example, define a predicate to be any string of words which yields a sentence when completed by one or more proper names. Thus '. . . is sitting three rows behind the grandfather of . . .' is a predicate by this definition, for inserting a proper name in each of the gaps yields a sentence, for example, 'Jane is sitting three rows behind the grandfather of Mary'. One simple way of defining a property would then be to say that to each predicate of English so defined there corresponds a property.

Such a definition would be useless for our purposes, for it fails to distinguish the genuine properties of an item from others. We would pre-theoretically expect that the changes in the properties of a thing are changes in that thing. Yet although I am now seated more than a mile away from my car, and although some time in the next hour I will be seated less than a mile away from my car, these changes will be taking place without any changes in me, or even in my position. But by our definitions just mooted, I have undergone a change in properties.

If we look into the history of philosophy, we find a more useful approach to the problem confronting us. For Aristotle, the properties of things were constitutive of their nature, responsible for their being the kind of things they are. Locke, with his sensitivity to the impact of science on our picture of the world, suggested that properties — what he called 'qualities' — were powers to produce ideas in us. Then he distinguished those powers which lay primarily in the objects themselves, from those which gave rise to various sensations and ideas in us but which were not direct correlates of the sensations thus aroused. For example, the colour we attribute to objects is in some sense 'in' them for they do have a number of primary qualities in virtue of which they seem coloured to us (distinctive surface texture, differential absorptions and reflection of light). The colour itself is, however, an artefact to some extent of the observational situation, and hence only a 'secondary' quality of the object.

The combination of Aristotle's and Locke's ideas suggests that we can think of genuine properties as related to causal powers possessed by objects (Shoemaker 1984: Essays 10 and 11). Shoemaker makes the interesting suggestion that the genuine properties of objects are high-level powers, powers to produce other, lower-level powers. The latter are the objects' normal powers to produce changes and events, while the former are conditional, for their exercise, on the presence of other powers. We can widen the Lockean point to the observation that these powers will not only cause responses in us, but will also be responsible for a range of responses involving other things as well. Not all the powers that an item possesses will be constitutive of its nature. A certain car has the power, let us imagine, to crush hedgehogs; yet such a power would not define it as a car, or as being the particular car it is. We can admit this even while noting the regrettable fact that cars are an important factor in hedgehog mortality.

The account of properties as clusters of powers gives us more or less what we need in order to maintain the thesis that unified, causally integrated whole objects are precisely those that have *genuine* properties *in their own right*. This is to say that they have powers and natures, and belong to sorts or kinds. The sum of the tree and the compost heap has no powers in its own right (although it inherits, so to speak, the powers of both the tree and the compost heap). In this way, the sum of the tree and the compost heap differs from the thermostat in Lovelock's electric oven. The thermostat has the power to maintain a preset temperature, a power it possesses in virtue of the relative configuration and causal powers of its parts. But the maintenance of the preset temperature is not some power it inherits from one or other of its parts. By contrast, the sum of the tree and the compost heap has the power to produce oxygen and reduce carbon dioxide by photosynthesis; but such a power is possessed by this artificial entity only because that is a power possessed by the tree and thus inherited by the whole entity from one of its parts. Although some of the properties of a unitary thing are typically inherited from its parts this is not true of the majority of what we might call its 'significant' properties. To explore this matter further, we would need to inquire into just what constitutes the *nature* of a unitary object, but I hope that it is now at least intuitively clear what the distinction between a genuinely unified thing and an artificial entity amounts to.

Powers, since they involve causal interaction with observers and other items, can be conceived as being genuinely relational. Notice, though, that there will be a difference between a genuine, relational property, and property which involves what Hume would have called a merely 'philosophical' relation. The latter kind of relation was involved in the example of my distance from the car, or the number of rows that one person is sitting behind another. Neither of these are genuinely relational properties. However, the tree's ability to photosynthesise is a power that involves essential relations to other things — in particular, to a source of light. Likewise, the power to subdue and digest prey is a genuine, but clearly relational, property of a predator. These observations raise a fairly tricky question, which will be discussed more fully in Chapter 8.

We are now in a position to provide a general answer to the question of whether the whole is more than the sum of its parts. We have seen that certain wholes, those that are unified and causally integrated, will be more than merely the sum of their parts, for they have properties — powers — which their parts lack, that is, properties that are not inherited from any part on its own. A mere sum of parts, by contrast, only has those powers which it inherits from its various parts alone, and has no powers in its own right. Notice that this point has been defended without recourse to doctrines of emergence, novelty, unpredictability, and so on that are often cited in discussions of this issue (Meehl and Sellars 1956; Nagel 1961: Ch. 11; and Klee 1984). In fact, the point is not specifically biological in any way. The observation just made is as true of ovens, cars and other artefacts as it is of the products of ecological and evolutionary processes.

Interestingly, the point is also independent of issues about reduction, and about ontological primitiveness. These two last issues, although looming large in some work on ecology, have already been shown to be independent of each other. The notion that some set of parts or particles is ultimately more real than everything else has been argued to be a hangover from the dubious degrees of reality doctrine found in Descartes' metaphysics. To be pro- or anti-reduction has also been seen to be of far less significance than some authors have thought. For questions about reduction in terminology have no direct existential or ontological import.

To argue that the whole is more than the sum of its parts, and to deny ontological privilege to any special class of things is still compatible with maintaining that the properties, or powers, of wholes are explicable, at least to some extent, in terms of the properties and interaction of their parts. There is nothing either terminologically or ontologically reductionist about taking this line on explanation. If there are such things as ecosystems and biological communities, then their properties will be explicable by reference to the properties and interactions of their parts. And it is to this question that we will turn in the next chapter, this one having cleared away, so I hope, much of the preliminary difficulties in getting to grips with that issue. There is no reason why we should not use the term 'emergent' to apply to properties displayed by wholes that are not simply inherited from their parts. Use of the term, however, should not be taken to imply that emergent properties are in some special way novel or unpredictable. Some of them may be so. But unpredictability would not be, as such, the hallmark of the emergent.

6.3 Nature and the natural

Since ecology is a branch of natural science, and is concerned with natural systems, we should take care not to let the notion of the *natural* confuse us overmuch. The term 'nature' is highly ambiguous and used by different people in a variety of ways. Given this variety, a case could be made for dropping the term from our descriptions of phenomena. Alternatively, we can aim at a minimal degree of clarity by distinguishing some of the senses of the term. In particular, it is easy to distinguish a broad, or absolute, notion of the natural, from a narrow, or relative, one.

When we describe certain kinds of human behaviour, or certain products or occurrences, as natural, we sometimes seem to have the following in mind. Human behaviour is natural when it is the sort of behaviour that can be found in other animals — particularly in other higher mammals. Thus it is behaviour which is common to humans and to other mammals. Although it makes the definition circular, we might add that we would want to exclude from such common, natural, behaviour any animal activity that has developed specifically in response to human interference. Products and events are natural, likewise, when their existence is not dependent on a certain kind of human management, production or interference. This is the broad meaning of the notion.

However unsatisfactory the definition, some examples show

that this sense of the term is what supports some uses of the words 'nature' and 'natural'. Ginseng root and willow bark are natural medicines, in this sense, while painkillers, stimulants and antibiotics synthesised in laboratories are not. A rat can chew a root or piece of bark as well as any human, but only intelligent, self-conscious, language-using organisms can build laboratories, brew their own chemicals and devise machinery for packing the ingredients in brightly coloured gelatine wrappings. As far as natural and unnatural behaviour is concerned, any student of animal behaviour will quickly assure us of the naturalness of homosexuality, the ingestion of psychodelic drugs, cannibalism, and fighting for leadership of the pack. At the same time, going to war, driving cars and reading books are unnatural activities in this sense of the term.

I think it is undoubtedly the case that we do use the terminology of 'nature' and 'natural' in this way some of the time. However, there is no immediate association between what is natural in the broad sense and what is morally right or good. There are those, of course, who urge us to adopt more natural lifestyles. But if they are using the term 'natural' in the absolute sense, it is not clear that we should pay them any heed. There is nothing morally better about the natural than about the unnatural. Leaving morals aside, there is nothing about the natural — in this sense — to commend it anyway. We are just as likely to be poisoned by natural as by synthetic substances and if ginseng and bran are good for us at all, it is not because they are 'natural' in this sense.

By contrast with this sense, there is another way of speaking about what is natural which may be of more relevance to considerations about what is right for us and what is good. There was a debate some years ago about the naturalness or otherwise of a vegetarian diet for human beings. Evidence for and against this idea was given by studying human dental and digestive arrangements, comparing these with the dentition and digestion of creatures known to be herbivorous and others known to be omnivorous or carnivorous. The idea behind these arguments was a simple one. We would find what diet was natural for us by discovering the kind of diet for which we are adapted. As Lorenz once argued, the horse's hoof mirrors the steppes, for we can read off, as it were, from a study of the hoof exactly the terrain for which it is adapted (Lorenz 1941). Likewise, we might say, our digestion 'mirrors' the kind of food that best suits us. Think of human beings sharing the same genetic recipes for certain general characteristics. The injunction to lead a natural life in the narrow, relative sense might then be interpreted as commanding us to lead a life that is close to the one for which these innate endowments equip us — a life in keeping with our nature.

Here we have a notion of the natural which is potentially significant for the purposes of determining what is good for us, and maybe even what is morally right. A natural diet may be one that — given our dentition, digestion and overall architecture — promotes health and fitness, and this could hardly be a bad thing.

We need to be wary of certain initial difficulties with this relative conception of the natural. *Human nature*, understood in this way, can characterise absolutely everything that humans ever do: there is a sense in which the most abominable wickedness, as well as the most sublime saintliness, can both be said to be expressions of a human nature. If, on the other hand, we restrict the sense of 'natural' to those dispositions and characteristics that are simply biologically innate, then not all that is programmed into us is either beneficial or morally worthwhile. In fact, we face here a problem akin to the one posed by Mill in his essay on nature. For he distinguished the sense of 'nature' in which the term refers to the entire system of things or events — in which case everything that happens is natural — from the sense of the term in which it characterises what takes place without the voluntary or intentional agency of humans. On the basis of this distinction, Mill was able to argue that the injunction to follow nature was either meaningless or pernicious (Mill 1910; also discussed in Millar 1988).

What we need to do to answer Mill's objections, thinking of them as focused on our relative conception (the idea of what is natural for humans) is to find some way of maintaining that the act of forming communities of value, or aspiring to be moral persons, or leading a life of virtue, is natural for us, while destructiveness, immorality and vice involve corruption or betrayal of our nature. What relevant evidence there is on this matter from biology and anthropology does not, I think, make very plausible the claim that a satisfying answer to Mill can be found. In what follows, therefore, I will not be arguing that any particular form of life is natural for us in the required sense. Nonetheless, it does seem that we do have, in this latest conception of the natural, something of more interest to the moral philosopher and theorist of human nature than our earlier, unrelativised notion.

If I decline to develop my views in terms of what is natural for us as human beings, it does not mean that I will have nothing to say about respect for the natural. If we return to the first proposed definition of 'nature', we can recognise the existence in the world of things and events that are independent of a certain kind of human agency and intention. Any that are the product of human action will have involved action that is natural in the sense of being common to humans and other mammals. This is not at all the same as Mill's notion of things being natural when they occur without the agency or intention of humans. On the earlier, absolute account of the natural, a voluntary act of human agency can be perfectly natural — so long as it is the sort of thing that a non-human creature also does. There are those who would say that eating, fighting and killing are different kinds of thing according to whether they are done by human or by non-human creatures. Although there is some truth in this claim — human actions, for example, are open to moral appraisal in the way that other animals' actions probably are not — there is not enough truth in it to pose an objection to my claim that some of our behaviour is to be thought of as natural, namely the behaviour we exhibit in virtue of our common mammalian or animal nature (Matthews 1978).

In this sense of the term, there is, I believe, some merit in the claim that the natural deserves our consideration and our respect. The experience of childbirth may be painful, unpleasant or akin to having surgery. But for some women, it is a profoundly moving and life-enhancing one. This is — I suspect — not unconnected with its naturalness. To feel one's body taken over in a way which leaves the agent with little intentional or voluntary control, to know that contractions will increase in intensity and come at ever decreasing intervals until the baby is born, must be to be caught in the grip of something that transcends the ordinary in experience. I have chosen this case as an antidote to the normal one, for advocates of nature, those who would have us respect and cherish things 'natural, wild and free' often point to the more remote experience of the natural — like Leopold meditating on the joys of wilderness (Leopold 1949). Childbirth is an especially striking example of the wildness within us, and it is only one of several cases where we can appreciate the natural at first hand, if only we would take the time to do so. Nothing I have said so far should be taken as suggesting that the natural is not something which is awe-inspiring and deserving of respect. Just what we should make of this, however, will have to wait till the later treatment of moral issues.

7

Ecological Explanation

7.1 Modes of explanation

The topic of explanation in the sciences is too large to tackle in a chapter like this. However, it is possible to clarify one or two puzzles that typically emerge in the study of ecological explanation. To begin with, I will follow Achinstein in distinguishing three different aspects of explanation (Achinstein 1983). First, there is the act of explaining; when this is carried out linguistically, as it often is, it will be an *illocutionary* act — an act carried out by a speaker or writer with the intention of producing a certain effect on an audience. The same sentence can be used in the execution of radically different illocutionary acts. Thus, if I say, 'The zero isotherm is at two thousand feet', I may simply be describing a fact about the local weather conditions. However, in certain circumstances, that same sentence could be partly constitutive of my act of warning a pilot of the likelihood of icing. Here, warning is an illocutionary act.

For an act of this sort to be successful, many different conditions must be satisfied. The fact that what I say is true is by no means the most important of these. Since such an act involves an intention on the part of the person undertaking it, one thing we can ask about the act of explaining is: with what *intention* is such an act carried out? Again following Achinstein, it seems reasonable to maintain that if I am to explain something to you, I must intend that what I say or write makes the thing in question understandable to you; it should also correctly answer some question that might have been put regarding the thing to be explained. As soon as we specify the act of explaining in this way, we are drawn to consider the nature of the questions that might have been asked about the item under consideration, and the nature of the 'instructions', as Achinstein calls them, that may be associated with appropriately scientific answers to these questions.

Without rehearsing the sort of detailed work that can be done on the basis of these few observations, it is possible to make a preliminary stab at characterising ecological explanations. For we can think of ecology as having a range of questions associated with it, namely questions concerned with the distribution, abundance and interactions of organisms. Suppose we want to know why a certain mature birch tree is 30 metres high. There are a number of different things we might be after in asking this question, but one understanding of it is as a question concerning the effects of the tree's environment on its size. Thus construed, the question is an ecological one, and any answers to it will be in the form of an ecological explanation.

We can think of the product of the act of explanation in terms that are to some extent independent of the act itself. This is the second of the three aspects of explanation identified by Achinstein. The product of the act is an explanation — a sentence or set of sentences written or uttered with explanatory force. In the sciences, there are fairly clear constraints on the content of such products. Not every set of true sentences relevant to the phenomenon to be explained will constitute an explanation of that phenomenon. Instead, the following are the kinds of constraints to be met before we are sure that the product is any sort of scientific explanation at all. The sentences may mention causal factors responsible for the phenomenon in question, or responsible for other things related in other ways to the phenomenon in question. Or they may deal with regularities associated with laws of nature — the phenomena to be explained thus falling under, though not being caused by, natural regularities. Or they may explain how certain results are derivable from other, already known, results. Or they may involve classification, showing that the properties or phenomena to be explained satisfy the defining conditions for being of the sort to which the explanation makes reference.

These rather general ways of making the distinctions among kinds of explanation can be clarified by running through a simple list of examples. Consider Gause's experiments on *Paramecium*, described in Chapter 4.2. *P. aurelia* when grown with *P. caudatum* always dominated to such an extent that the latter was rendered extinct. One causal explanation of the extinction of *P. caudatum* would involve the claim that the cause of its dying out was the

consumption of its essential resources by *P. aurelia*. By contrast, a
non-causal, but perfectly scientific explanation, of the moon's
effect on the tides, is to say that the moon's gravitational effect on
the waters of the earth is a special case of the law that all bodies
exert gravitational forces on other bodies. The law of universal
gravitational attraction does not itself *cause* the moon's effect on
the earth. So subsumption under a natural regularity is a form of
explanation that makes no explicit appeal to causation. (The
example of moon and tides is borrowed from Chapter 7 of
Achinstein 1983).

For an explanation in the form of a derivation, consider the
Lotka-Volterra equations, described in Chapter 4.2. These equations apply to resource utilisation for two species. Suppose, now,
that we wish to consider the situation of three species, trying to
determine the carrying capacities which would allow an
equilibrium among those species and various degrees of niche
overlap. What we would do is try to derive a three-species model
from the two-species one by making a number of assumptions
about the shape of the resource utilisation curves for the three
species and expressing the competition coefficient in some suitable
formulae (to see how this may be done, look at Chapter 7.5 of
Begon, Harper and Townsend 1986). Again, notice that although
there may be causal factors in the background, the explanation
makes no explicit reference to causation: we are not claiming that
the two-species equations — or anything else — cause the
derivability of the three-species case. Finally, for a simple example
of classification explanation, consider the explanation of why a
particular species of cynipid wasp is a parasite. Suppose this is a
wasp which lays eggs in oak trees. The response of the oak, in
producing galls, shows that the wasp is a parasite rather than an
item that stands in a weaker, commensalist relation with the trees.
Of course, the eggs cause galls to form. But the classification of
the wasp as a parasite depends on its provoking a response from
the host: that the wasp produces such a response from the oaks
does not *cause* it to be parasite, but certainly explains why we
classify it as one.

There are other kinds of explanation as well, and it would be
unduly laborious to try to go through them all. The important
point to bear in mind is that not all scientific explanations involve
explicit appeals to causality — they are thus not all causal explanations. This is not to deny that the sciences are concerned with the
causes of things — for they certainly are. It is simply to point out

that, even where our primary concern is with why things happen the way they do, not every kind of explanation will involve explicit citations of causal factors.

The third aspect of explanation mentioned by Achinstein is the evaluation of the product of explanatory acts. Not all explanations are equally good and, however careful our definition of 'explanation', there will be many things that fit such a definition yet seem hardly worthy of the title at all. A great deal of effort has been expended by philosophers of science in trying to specify just what makes an explanation good, scientific, genuinely explanatory, and so on. In fact, I am doubtful if any of this work has yet put us in a position to make appropriately general pronouncements about the quality of explanation. A good guide, here as elsewhere, is to take the actual practice of leading scientific practitioners as — by and large — a clear example of decent explanation. If such examples happen to fit models that specify falsifiability, exposure to potential disconfirmation and so on as features of good explanation, then so much the better.

One particularly well known model of scientific explanation, especially of the causal sort, is Hempel's development of Hume (Hempel 1965: Essay 10). According to Hempel, we are in possession of an explanation of sentence p when we can show that p follows from some other sentence by virtue of some covering law. In other words, a good explanation has the form of an argument which is deductively valid and whose conclusion follows from the premises by virtue of the use of natural laws as premisses. Thus, the following would, by Hempel's lights, be a good example of scientific explanation:

All plants have higher carbon-to-nitrogen ratios than insects; *Rubus idaeus* is a plant while *Byturus tomentosus* is an insect; *Rubus idaeus* has a higher carbon-to-nitrogen ratio than *Byturus tomentosus*.

The last sentence here is explained by virtue of the fact that the law-like statement (the first one) ensures that it follows, by deductive logic, from the second one. Moreover, given the first two sentences, we could deduce the third one as a prediction about what we will find if we scrutinise the chemical constitution of members of the appropriate species. Thus prediction and explanation seem to be nicely symmetric.

Obviously, the Hempel scheme, shown here only in its simplest form, allows for the possibility that explanation can go on and on. There is no reason why we should not produce similar schemes that would explain why plants have higher carbon-to-nitrogen ratios than insects. Notice that although the Hempel scheme is well suited to cases of causal explanation, the example just given is not causal — for the properties of being a plant and being an insect, respectively, do not cause the different ratios in the species. Finally, notice also that the scheme allows for additions concerning falsifiability, testability and the like. Thus, we might add that the covering law in any proper explanation must be falsifiable, and that if the explanation is to have any plausibility, the law should have resisted deliberate attempts to render it false (Popper 1959, 1963).

Regrettably, as is well known, there are strict limits to the enlightenment we can acquire by adhering to the Hempel scheme, even with the additions just mentioned. Among the problems it faces are that many examples which fit the scheme look like only the thinnest of explanations — even supposing it right to call them 'explanations' at all. This problem is connected with the general problem of law — what it is that distinguishes those general statements that are laws from others that are not laws. Suppose that Mary always brings life, humour and happiness to any party she attends. You know this; so do I. Then, if you ask me why the party at Foot's place last Saturday was such a success, I can give a laconic explanation by telling you that Mary was at it. Given what we both know, such an explanation plainly suffices. But now suppose that the background story is rather different. This time, we both know that Mary is a quiet person, not given to socialising, but somehow with a knack of attending only those parties which are a success. If, with this as background, you now ask me why last Saturday's party was such a success, I do not give any explanation of this by telling you that Mary was at it. Of course, if you did not know whether the party was a success or not, Mary's skill at identifying the good ones would have given you a reason for thinking that it was a good party. But only in the former, not in the latter, case would you have an explanation (Wilson 1979; I have benefited also from access to Wilson's unpublished work on conditionals).

If we read these results back into the Hempel scheme, we find that the same scheme underlies both stories, even though there was an act of explanation in only one of them. In the second case, the instance of the scheme is something like this:

Any party that Mary attends is a good one;
Mary was at the party last Saturday;
The party last Saturday was a good one.

The trouble here seems to be that the general statement about Mary's attendance at parties can itself be more or less explanatory. Yet the whole point of the Hempel model, and of other related models in the philosophy of science, is to give a non-circular account of what makes a good explanation. It will not do to claim that such explanations must involve explanatory generalisations. But unless we say this, the schemes like the one above will, while fitting some cases of scientific explanation, allow us to count as explanations things that are quite clearly not explanations at all (Achinstein 1983: Ch. 4; also Hempel 1965; Salmon 1971; and Brody 1972).

As far as the philosophy of ecology is concerned, the problems about explanation discerned in the present chapter will apply to those cases that represent answers to specifically ecological 'why' questions. Some work in ecology has certainly not been guided by appropriate deference to the ideals of testability and falsifiability, although awareness of these aspects of scientific methodology seems fairly widespread these days (Strong, Simberloff, Abele and Thistle 1984: Preface, and Introduction by May). But ignorance of that in which a phenomenon consists has never been a deterrent to discussing it either in philosophy or in the sciences. So in the following sections, ecological explanation will be discussed, along with the nature of the concepts and entities that figure in such explanations.

7.2 Theories and things

There is a peculiar difficulty about deciding just what the focus of ecological explanation is. The key terms of ecology — terms like 'community', 'niche', 'predator', 'resource', 'system', 'competition', 'parasite', 'detritivore' and so on — do not have one clear level of application. From one point of view an individual plant is a food resource for dozens of populations that may inhabit it; from another point of view it is a population with variable resistance to disease and predation; and from yet another it is but a minute part of a larger community with a small role in the overall chemical cycling of the community as a whole. This lack of focus has led ecology to fragment, to a certain extent. There are thus community ecologists, population ecologists, systems ecologists as well as plant and animal ecologists who focus their studies on specific interactions within a community.

In this way, ecological studies have a breadth of scope and application that is not typical of the physical sciences. There was a time, certainly, when it seemed that the mechanics of planetary systems would suffice for explaining the behaviour of the atom and, as we noted in Chapter 5, there is a sense in which thermodynamics is no more than statistical mechanics. But these are trivial specimens of comprehensiveness compared to the case in ecology. Ecology, in fact, is a potentially good example of a science in which the theoretical apparatus, its laws, concepts and terms, can be shown to be quite separate from the observational or experimental results in the field. If the variability of a population, for example, means that it will show a differential rate of succumbing to disease, then we can expect to observe this phenomenon at the level of the individual tree, and at the level of the worldwide distribution of a species, and at various levels between these two.

This flexibility in the application of theoretical notions to observed phenomena is no weakness in a science. The meaning of theoretical terms is given, at least in part, by the role these terms play within the theory. Theoretical ecology, thus, will give at least partial guidance in the understanding of what a *community* is, or what a *niche* is (recall the definition of the latter term as a volume in hyperspace). We can then look for observational or experimental examples of items that form a community or occupy a niche. Some of these cases may, upon study, force us back to a new refinement in the theoretical notions, but this is no more than the usual interplay of theory, observation and experiment.

Likewise, even in the physical sciences, we will find the same sort of interplay between the theoretical and the experimental. One well known example concerns the charge on the electron. When Millikan helped establish, using his famous oil-drop experiment, that all electric charges are multiples of a certain elementary charge e, this gave a means of maintaining that the charge on the electron was this same, elementary charge (for details see Chapter 5 of Nagel 1961). Yet the oil-drop experiment in no way determines just what it is for something to be an electron. The notion of electron is defined, at least partially, by various atomic theories, and there is no reason why Millikan's determination of its charge should not be compatible with very widely differing accounts of just what this thing is. Indeed, the history of physics in the present century has seen the production of several such accounts.

This quick demonstration of the semi-independence of theory and observation is not meant to show that any old theory is compatible with any set of observations. The skill of the scientist involves the recognition that certain kinds of observation commend certain sorts of revision to theories and hypotheses under test. There is, of course, a philosophical possibility here, originally described by Duhem and more recently supported by Quine. This is that any hypothesis can be retained in the face of adverse observations, provided we make enough adjustments to the rest of our theory (Duhem 1914). But from the range of retainable hypotheses, the working scientist faced with adverse observations retains some and discards others.

To show the kind of problem faced by ecologists in this respect, consider the nature of explanation that might be undertaken for two very different kinds of community. Think, first of all, of a tree. Now, individual trees are distinguished from other trees by the fact, let us say, that each occupies a connected and distinct volume of space. At no time does the volume of space occupied by one tree overlap with that occupied by another. Yet trees are modular organisms. That means that a tree results from the development of a zygote which first of all produces a unit of construction out of which other units of construction similar to the first are produced. We recognise such modular organisms, in the main, because they take a branching form of development and are usually non-motile. Their shape, size and overall form are highly dependent on environmental conditions (compare a mature oak in a broad leaved forest with a bonsai specimen in a terrarium). By contrast, many animals are unitary organisms which adopt a determinate, highly structured, pre-programmed form which, with due allowance for growth, damage and senescence, they maintain throughout their lives (Begon, Harper and Townsend 1984: Ch. 4.2).

Modular organisms pose an immediate problem for ecologists when resource abundance is under consideration. Whereas the number of small mammals available for larger predators or raptors in a certain habitat is (at least in principle) determinable by counting the individual mammals, no such procedures will work for a modular organism. For instance, water hyacinths and water lettuces just fall to pieces — quite literally — as they grow. So one original zygote can be responsible for the entire population of water hyacinths in a pond or lake — they are all simply modules of the same original, and thus constitute one genetic individual. So a measure of resources may have to be obtained by counting the number of available modules in a habitat at a time, rather than the number of genetically distinct organisms.

Clearly, then, there is a way in which we can think of a modular organism, say a tree for simplicity, as a community of modules. As a community, the organism in question may constitute a variable resource for predators. For example, suppose that after caterpillar attacks on one part of the tree, neighbouring leaf modules begin to secrete a substance that is poisonous to the attackers. Whereas some of the tree was a usable resource for the caterpillars originally hatched there, neighbouring parts may be of no use to further populations of the same caterpillar that simply happen to have hatched slightly later. But what is true of predation is also true of disease. When struck by various diseases, an individual tree is likely to show variable resistance to the attack. Its modules form a community of variable resistance — perhaps due to slight genetic variability from one module to the next (Whitham, Williams and Robinson 1984).

It seems clear in a case like this that the tree has properties and features in its own right. Its own variable response to attack, whether from disease or predation, is a biologically interesting feature, even though this feature can be described in terms of ratios of affected-to-unaffected modules. We have no difficulty in comprehending this kind of observation because the community of modules in question is coextensive with a recognisable individual item. What goes for disease and predation also applies, for example, to the geometry of leaf orientation, where we can discern rules that have some success in predicting the likely shape of the tree given its position in a forest. What is more difficult is the recognition of community-level features for communities that happen not to coincide with recognisable individual things. Yet if ecology is to say anything interesting about distribution and abundance, it must deal in community-level descriptions and predictions, that is descriptions and predictions of ratios and relative abundance of populations in a community. In a loose and misleading way of speaking, we could say that ecology is committed to the reality of communities, or the recognition of such entities as communities.

In order to find out just what is involved in the appropriate theoretical recognition of biological communities, I will consider in the following section, a case that threatens to cast doubt on the notion that communities, or ecosystems, are things of any biological significance at all. This case purports to show that certain trends in the development of forests are independent of any biological features that forests may possess in their own right. Although a tempting reply to this kind of case is simply to admit that forests are neither ecosystems nor communities, this reply is shown to be too shallow. Rather, we have to distinguish explanations at different levels, taking care not to confuse the causal explanation of a particular occurrence with the explanation of why certain communities show the distribution of populations that they do.

7.3 Forest succession

As noted in Chapter 4, a feature of traditional ecology was the emphasis on the ordered progression of a community or plant formation through a number of phases in which different populations were present. These several phases of a succession were supposed to lead to a small number of determinate outcomes, so-called *climax* communities. The idea was that the climax represented a stage of mature stability for the community which would persist. Alternatively, the climax might itself have a cyclic form, in which one pattern of community replaces one or more other patterns, which recur rather like a repeated check in chess. Finally, to deal with apparent counterexamples to this account of succession, the notion of an *arrested climax* gave a kind of *ad hoc* account that covered cases where the progression towards the climax state had stopped at some early stage (Putman and Wratten 1984: Ch. 4).

To deep ecologists and those attracted to the deep position, the climax community is sometimes presented as the natural fulfilment of ecological and evolutionary processes; by 'natural' here is meant 'undisturbed by human interference'. Despite Clements' observation, quoted earlier, that grassland and forest are the usual outcomes of succession, many have added other features to the characterisation of these climax communities: they are to be thought of as biologically rich, diverse, self-sustaining, an end or final cause, in Aristotle's sense, towards which all living things have a natural movement (Clark 1983). There is, alas, little biological or ecological credence to be placed on such an ideal and, as we will soon see, little evidence that natural systems are really moving towards any such ends.

However, the general conception of succession towards a climax stage can be divorced from the ideals of the deep ecologist. It certainly makes good ecological sense to see if natural communities do move through discriminable phases, and whether any biological principles, hypotheses or laws may apply to such movement. The case to be considered in the present section deals with precisely such a question. For it concerns the changing distribution of trees in a typical New England forest. In terms of our discussion of evolution and ecology in Chapter 3, the issues to be discussed in connection with the case may seem to be at least as much evolutionary as ecological. However, in what follows we will find problems about natural selection which parallel those which arise in connection with explanations in ecology.

In order to understand the research I am about to describe, some grasp of stochastic theory and of Markov processes is required. If we think of a series of happenings over time, and the probability of the occurrence of one event given that certain others have occurred, then we are engaged in *stochastic* thinking. Markov processes are one very special sort of time series. Suppose that we are interested in events e_1, e_2, e_3 and so on up to some event e_{n+1}, each of these events being equally spaced. If the conditional probability of e_{n+1} given e_n is exactly the same as that of e_{n+1} given *all* the events in the time series up to e_n, then the process in question is a Markov one. Quite simply, then, Markovian systems are ones without memory — where only the immediately preceding state of the system is relevant to the probability of the given state.

In studies of woodland, Horn wondered if Markovian reasoning might be used to predict the outcome of competition among various tree species. Examining the pattern of competition among species showed that two things could be discovered quite straightforwardly. For a 50 year period, let us say, it would be possible to discover the percentage of trees of a given species in a habitat that are still standing. As well as this survival rate for individual trees, it would also be perfectly possible to plot the ratio of replacement for those individuals that do not survive the full 50 years. For the latter, it will be possible to calculate how many have been replaced by other trees of the same species.

This information on survival and replacement rates can be

formulated as a *life table* for the tree populations. Suppose, to take a simplified, fictitious case, that we investigate populations of just four species in a New Jersey wood — grey birch, black gum, red maple and beech. Of the original grey birch population now observed, let us suppose that in 50 years (when the next observation is made), only five per cent are still standing, 36 per cent have been replaced by black gum, 50 per cent by red maple and nine per cent by beech. For black gum, the 50 year figures are rather different, 37 per cent are still standing, 20 per cent have been replaced by further black gum trees, while one per cent have been replaced by grey birch, 25 per cent by red maple and 17 per cent by beech. In this way we can start to draw up a life table showing the fortunes of our four species over 50 years (Table 7.1). Let us now suppose that this 50 year cycle repeats indefinitely. No matter what distribution of the four species we start with, the long-term outcome will always be the same. What will happen, in the absence of any disturbance, is that the beech will predominate — hardly surprising, given its longevity and high replacement ratio — grey birch will become extinct, and the two remaining species will hang on in small numbers. In Table 7.2, the result of applying these

Table 7.1: Horn's Markov matrix, from Horn 1976

Now	50 years hence			
	Grey birch	Black gum	Red maple	Beech
Grey birch	5 + 0	36	50	9
Black gum	1	37 + 20	25	17
Red maple	0	14	37 + 18	31
Beech	0	1	3	61 + 35

Diagonal is % of trees still standing plus % that have been replaced in 50 years by another of their own kind. Off diagonal terms are % of trees replaced by another species in 50 years. The % of trees still standing was estimated by assuming that trees die at a constant rate such that 5% are left standing after 50 years for grey birch, 150 years for black gum and red maple, and 300 years for beech.

Table 7.2: Successional convergence over time, after Horn 1976

Age of forest (years)	0	50	100	150	200	. . . ∞	Very old forest
Grey birch	100	5	1	0	0	0	0
Black gum	0	36	29	23	18	5	3
Red maple	0	50	39	30	24	9	4
Beech	0	9	31	47	58	86	93

calculations in the case where we start with a wood consisting entirely of grey birch is shown. The eventual outcome is compared with actual tree counts for the species concerned in old, undisturbed patches of New England forest. It seems clear that the stochastic model is confirmed to some degree by these observations.

Horn's results are interesting for a number of reasons. One is that they show that a successional process need not result in a situation of stable diversity. For in the case of our trees, we have stability, exclusion of one species and perilously small numbers of the other two. Of course, natural forests will sometimes display diversity precisely because they are disturbed by factors such as fire, weather damage and human manipulation. Those deep ecologists who favour diversity might do well to dwell on this point, for there will likely be cases where the sort of diversity of species that is pleasing to us can only be maintained by active management on our part.

This aside, Horn's results suggest something of more consequence still. What has happened to our imagined tree community over time is that the distribution of species within it has changed. But has there been any biological significance to this change? Horn thinks not, for he writes, 'several properties of succession are direct statistical consequences of a species-by-species replacement process and have no uniquely biological basis' (Horn 1976: 196). The suggestion is that succession is not so much a biological process as a statistical one. Let us agree that there are ecological principles governing interspecific competition — a point that is supported by much current work in ecological theory. Even so, it can be argued, Horn's results show that succession is simply the outcome of competition at the level of individual trees or, at most, at the level of populations. The whole forest community has still no biologically significant role in the process.

I think it is helpful here to contrast happenings and patterns that are of biological significance with the issue of what items play biological roles. The death of one grey birch and its replacement by a red maple is not a process in which the forest plays any part. But that the forest changes through time in the way shown on the tables may nonetheless be a matter of real biological interest. That these changes over time are Markovian in character is also of interest, just as it would be of interest to find that we could get the same results by throwing dice. In short, there need be no opposition between the biological and the statistical. That a community changes in a random way, to take the most extreme case, may well be something of considerable biological significance. Moreover, the changes are facts about the community, and so there is a community-level explanation of change to be given in Markovian terms.

We can try to confirm this claim by contrasting this case with a similar problem that arises in evolutionary explanation. It is a trite observation that, given the Darwinian theory, no organism evolves. This is in contrast to one understanding of Lamarck's theory, where the drive of evolution is supposedly supplied by slight, but cumulative changes in organisms over their histories. Given the transmission of these slight modifications to their offspring, whole species could be said to evolve in virtue of the changes in their members. Evolution, understood in this way, is not really a fact about populations, species or communities. It is, rather, the consequence of individual development aggregating over time.

Darwin's account, on the other hand, simply requires that some traits present in parents will reappear in their offspring, that is that they are *heritable*. Given heritability, we do not need to imagine any transmission of developed traits, or even any development at all, on the part of organisms in a species. For evolution will simply consist of a changing distribution of properties across populations. What natural selection will explain is the ratio or frequency of various properties over a population. Traits, or properties, that reduce an individual's fitness will tend to reduce in frequency over time, while traits that increase fitness will increase in frequency. It is not, however, natural selection that kills an individual organism before it reaches reproductive age: it is the parasite it has no defence against, or the weather for which it is ill-protected, or the members of its own or other species which outcompete it in the struggle for resources, and so on. Put crudely, natural selection is not a causal factor in explaining the traits, fitness or survival of given organisms, even though these things all have a bearing on selection. Rather, natural selection will account for why physical and behavioural traits have the frequency they have in a given population (Sober 1984b: Ch. 5).

Sober, whose account of selection I have been following here, points out that Darwinian theories of selection are not to be understood as theories about change in objects. We can make the objects upon which selection works as static as we please, and yet still give a Darwinian account of the changing nature of the population we are studying. A simple analogy that he uses shows clearly the difference between developmental and selectional explanations (Sober 1984b: 149). Suppose we are confronted with a room in which all the children are at a certain stage in their reading. The *developmental* explanation of this phenomenon proceeds by showing how each child in turn has developed from an earlier reading stage to its present skill level. By contrast, a *selectional* explanation may suggest that, in some other room, there are candidates for admission to the one under study. Some read at a certain level, others do not, and only those who read at the appropriate level are admitted to the room.

Now think of explaining the length of giraffes' necks. On the developmental account, we think of aggregating changes from giraffes with shorter necks up to giraffes with long necks and with not-so-long necks. Survival to reproductive age, like admission to the room, selects those with a certain neck length as constitutive of the breeding population. Now, on the latter account, what causes a given giraffe to have a long neck is not natural selection. Rather, it is its genetic endowment from its parents. Sober concludes that what natural selection explains is an irreducibly population-level fact rather than features of individuals.

Now let us compare the two cases — natural selection as described by Sober, and the earlier results about trees. Each case concerns the distribution of characteristics across populations or communities. Just as Sober maintains that we do not explain the length of the giraffe's neck by appeal to selection, so we could not explain the demise of the last grey birch by appeal to Markov processes. The giraffe has the length of neck it has because it inherited certain genes for long-neckedness from its parents. The last grey birch died because it was struck, let us suppose, by lightning and was then attacked by a lethal fungus. Yet natural selection explains the neck length of giraffes, and Markovian processes explain the distribution of species in our imagined wood. If natural selection is a biologically interesting phenomenon, it follows that Markovian processes in succession may likewise be of biological interest. Moreover, the Markovian process applies to the stages of the wood or forest, not to the individual trees. So if there are population-level explanations in evolutionary theory, there are likewise community-level explanations in ecological theory.

There are a number of reasons for being dissatisfied with the argument just given; some I will discuss, though not all, in the following section. A real problem that still confronts us is establishing just what is to count as a community, how best to describe community structure and how to understand the kind of quasi-functional explanation that can be given of the roles of items within communities. Leaving this particular problem to the next chapter, let me conclude this one by looking at a necessary qualification that Sober's view seems to require. Suppose, to take another of his examples, the cause of my having opposable thumbs is that the genes inherited from my parents contained instructions for so building my hands. Thanks to natural selection, however, all the individuals in a given population have, let us suppose, just such opposable thumbs. On Sober's account, the frequency of traits in a population can be explained by natural selection even though individuals' possession of these traits cannot. But to say this is, in my view, to be unduly restrictive in understanding explanation.

No doubt the proximate cause of my possession of opposable thumbs is the one just mentioned. But why is my gene complement the way it is? If there are genes for opposable thumbs in every member of the population under study, then one explanation for my having such genes is the general truth that everyone has such genes. If natural selection explains the presence of such genes in *every* member of the population, then it must explain *my* possession of such genes. What is wrong is that Sober takes too causal a view of explanation. As we have seen, not all explanations in the sciences are causal. The causal fuel that drives natural selection will be the genetic mechanism of inheritance and the ecological interactions between organism and environment; natural selection will thus be a causal consequence of these 'bottom up' features. But this does not rule out all kinds of 'top down' explanations, many of which will not be causal.

Thus, in the case of Horn's forest, we can likewise distinguish the causal factors that drive the Markovian system from the possibility of explaining the demise of the last grey birch not only in the way suggested but also by appeal to the successional process itself. Of course, the features described in Tables 7.1 and 7.2 do not cause the extinction of the grey birch population. But they may well feature in an ecological explanation of the population's demise, and such an explanation must have some relevance to explaining the demise of the last individual belonging to that population. The last individual was destroyed by the factors mentioned. If we are to explain the demise of the last of the population, then a complete explanation will tell us why it was the last (why there were not others) and the Markovian features of the system are relevant to such an explanation.

It may seem odd that I have accused Sober of holding that all explanation is causal given his explicit denial that this is so. However, as I read him, his denial is more apparent than real. The key passage, where he contrasts causal explanation with what he calls 'equilibrium' explanation, goes as follows:

> Equilibrium explanation shows why the actual cause of an event is, in a sense, explanatorily *irrelevant*. It shows that the identity of the actual cause doesn't matter, as long as it is one of a set of possibilities of a certain kind. Disjunctive properties do not cause, but they can explain. (Sober 1984b: 140)

The idea is this. Forces like mutation, migration, genetic drift and the like give rise to successive gene frequencies in populations. Now the specific forces in the actual history of a population may be unknown, as may be the actual past gene frequencies. But this does not matter, for what evolutionary models allow us to explain is the occurrence of a present genetic frequency, whatever the causal route to it may have been. The equilibrium explanation covers several causal cases, while appealing to no specific one. But — although Sober is right to say that such an explanation does not cite a cause — the explanation is still causal at one remove. This he recognises himself when he writes that both causal and equilibrium explanations

> explain an event by showing how it is 'embedded in a causal structure'. They differ insofar as the former emphasises the actual trajectory of the population whereas the latter shows how the event to be explained could have resulted in a number of possible ways. (Sober 1984b: 142)

I conclude, therefore that recognition of the category of equilibrium explanations is not enough to reduce the impact of my challenge to Sober's view. My claim is that there are non-causal explanations of a biologically interesting kind, to be given at population and community levels. This claim, it should now be clear, is both similar to, but importantly different from, Sober's.

8

Supervenience and Essence

8.1 Functions and supervenience

There are two apparently incompatible ideas that spring to mind when we think about organisms in their relations to each other and to the environment. First, it seems clear that an individual organism, although the product of some kind of sexual, vegetative or other reproduction is not itself created with any specific role or function. Indeed, the parts of natural objects in general have functions, while they themselves, viewed as wholes, have no functions at all.

A second thought, however, that seems as convincing as the first, is that natural selection and environmental conditions have fitted organisms, whether individually or in populations, for certain niches within larger biotic and abiotic structures. If we look at an ecological system as a whole, we find it hard not to suppose that some of the populations and individuals within it are playing functional roles, which can be described in terms of their contributions to the maintenance and development of the system in question. It matters little, in considering this second thought, whether we focus our attention on ecosystems — communities along with their abiotic surroundings — or on the communities of living things themselves. On the latter perspective, more popular these days among some ecologists than ecosystems thinking, we recognise that abiotic items play important roles in community structure.

Not every reader will feel puzzled by the two views just mooted. For those of an Aristotelian frame of mind, it may seem obvious that each living thing has certain aims, goals or ends towards which it strives. Unless animals all shared the goal of staying alive and reproducing, it seems hard to explain common behavioural features — like feeding, competing and challenging for resources and mates, avoiding danger, and so on. Now, although there is a distinction to be drawn between goals or ends, on the one hand, and functions of items, on the other, there is also an easy way of connecting the two sets of concepts. For if an organism has a certain end, then the things it does in order to achieve that end will be among its functions. Shedding leaves in the autumn will thus be a function of a deciduous plant, if it is a means to achieving continuance of the plant's existence (by resource conservation).

But the fact that certain bits of behaviours have functions, just as some parts of organisms have functions, does not yet show that the organism as a whole has a function or set of functions. Indeed, there are grounds for claiming that, even though animals and plants can take on, and be given, roles and functions, yet they have no functions in their own right — they are intrinsically functionless. (I made heavy weather of these distinctions in Brennan 1984.) Although the shrub at the bottom of the garden hides the window of the neighbouring house, and functions to do so, that is not its function. Although a pair of raptors control the population of voles in a certain habitat, and again can be said to function to do so, doing so is not their function. The locution that something *functions* to do such and such, is perhaps misleading. For we very often say this simply to describe what some item in a certain situation typically does. The function of the dog's tail is to act as a rudder and as a signalling device, but his tail may also function to sweep the kitchen floor when he happens to be both pleased and seated there. Wright gives the striking example of the second hand of a watch. Its function is to measure elapsed time in seconds, but it may also, let us imagine, function to sweep away accumulated dust from the watch face.

However, let us now take up what Dennett calls the *design stance*. We think of a biological community as consisting of a number of populations interlinked by their use of resources — where, of course, one population may itself be a resource or producer of resources for one or more of the others. Let us imagine that over time the community reveals a 'steady' state: the number of certain kinds of predators may not stay constant, but it varies along with variations in the supply of preferred prey. The numbers of prey vary along with variations in other resources. Within the community there are certain constants. Matter cycles show the same pattern in lean and rich years, detritus is well used by detritivores, minerals circulate in almost a closed cycle, nutrients are highly conserved, and so on. Removing one or two populations, let us further suppose, would not destroy the community but would lead to the development of distinctive new energy, nutrient and mineral cycles. The whole ecosystem containing the community would move to a distinct, new state.

Ignore for the moment the empirical question of whether there are such structured, well organised communities in nature. Just suppose, for the moment, that there are. Now consider the thought that such a community might have been designed to operate as a device for using solar energy for cycling chemical agents, thus supporting individual life of various kinds of complexity, with each population playing its role in the story. From this design perspective, it seems to make good sense to think of the various populations as having functions — the detritivores function to make certain chemicals available for other members of the community, the raptors function to keep the population of their prey in check and so on. Just see the community from this new angle, then, and talk of function no longer seems widely inappropriate.

The solution to the difficulty is, I think, very simple. Of course, no-one designed the communities described, but the very fact that we can think about them as if they might have been the product of design makes functional description appropriate. But if it is appropriate to describe the population of lions in a certain habitat as having the function of controlling the numbers of gazelle, then it is equally appropriate to describe an individual lioness in that habitat as having the control of gazelle numbers as one of her functions. (This is denied in Brennan 1984, where functional roles were assigned to populations, with individuals being declared to be intrinsically functionless. The suggestion about to be made in this section seems to me to be a better solution to the problem than the one I gave previously.) What we can do, in fact, is admit that the lioness does have this function in the community under consideration and yet deny that such a function is one of her intrinsic properties. We can thus preserve the distinction between the intrinsically functionless, autonomous natural creature, on the one hand, and the truth of the claim that some creatures do have an ecological, or community, function on the others. Doing this consists in recognising the difference between what we might call the *basic* properties of things and their *supervenient* properties.

111

Although the notion of supervenience is widely used and discussed in the philosophy of mind and action, it is helpful, for our purposes, to consider simple kinds of supervenience. Think, for example, of the relation between the colour of an object and its properties of fine surface texture. We commonly suppose that there is a quite determinate relation between these two lots of characteristics. Although it may be that varying surface textures and materials can underlie the same colour, two items with the same surface material and texture will be just the same colour. Conversely, if an item changes in colour, then it must have undergone some change in the basic properties of its surface. On the face of it, supervenience claims look fairly uninteresting. To say that an item's colour supervenes on the nature of its surface seems simply to express the claim that its colour is a consequence — perhaps due to natural laws — of its having a certain kind of surface. However uninteresting such an observation may seem, we can develop it in the case of biological communities in such a way as to solve the problem stated at the start of this section. For it will turn out that organisms have two lots of properties: first, their basic morphological or taxonomic properties, relative to which we can classify them as members of biological kinds (more or less); and, second, their functional properties, some of which can be displayed by ecological investigation. The latter properties are supervenient on their basic ones given their location in communities of certain sorts. It will follow that taxonomically and morphologically distinct organisms can have the same, superve- nient, functional properties (the same ecological role), while a change in an organism's functional properties will mean a change either in its basic properties, or a change in its environing circumstances.

8.2 Ecological traits

Given the simple account of supervenience just mooted, it can be easily shown that many of the characteristics of organisms are supervenient on two other kinds of characteristic. First, there are the basic, or intrinsic characteristics of the items themselves: in animals — though not in plants — such characteristics include morphological or structural features, possession of organs of a distinctive sort organised according to set principles of physiological architecture. There is also the genetic program of the organism, its modes of response to various forms of stimulation and its instincts. Let us concentrate, for the moment, on morphology. Given our ability to classify most creatures by species, the most significant of their structural features will figure in the description of typical members of the species. Thus it is one of the intrinsic features of the leopard seal that it has large, cusped teeth. Since seals are a good example of creatures with fairly specialised dental arrangements, it is not surprising that other species of seal possess quite different kinds of dentition. To say that an individual seal belongs to a particular species — say the species *Hydrurga leptonyx* — is to say that among its basic, or intrinsic properties will be the dental architecture typical of the species.

But there is a second kind of characteristic to be considered before we can detail the supervenient characteristics that are of interest to ecology. The teeth of the leopard seal enable it to grasp, subdue and devour large animals, like penguins and other seals. In this way, it contrasts with the crabeater seal, whose teeth suit its habits of feeding almost exclusively on krill (Begon, Harper and Townsend 1986: Ch. 1.4). But the crabeater seal cannot feed on krill, nor the leopard seal on penguin, unless their environment contains krill or penguin. Given its teeth, its instincts and its other anatomical arrangements, and given the presence of penguins in the environment, the leopard seal is a penguin eater. The property of preying on penguins is thus supervenient both on internal, structural features of an animal and on its environment.

In fact, the leopard seal is only a seasonal penguin feeder. At other times, it consumes krill, other invertebrates, fish and, as already noted, other seals. What it feeds on, then, is just as dependent on what is available in the environment at a given time as on features of its anatomy. The ecological study of predator–prey relations states that the predatory habits of creatures are supervenient on characteristics internal to the creatures in question and also on characteristics of the environment. An obvious way in which changes in either of these will change the roles an animal plays in an ecological community is suggested by the findings of Jim Corbett, the famous tiger hunter. As he grew older, Corbett gave up shooting animals for sport, but continued to destroy maneating tigers and leopards. In all the cases of maneaters he describes, he found plausible reasons, due to disease or injury affecting the animals concerned, why humans and cattle had become the regular prey of these animals (Corbett 1960). Plausible though Corbett's findings undoubtedly are, it is also worth noting that the incidence of tigers preying on cattle and humans also coincided with a reduction at the same time in the normal prey available to them as a result of destruction of large areas of habitat. It may well be that both disease or injury and environmental factors led to the appearance of so many maneaters in Kumaon between the wars.

Some of the examples quoted in Chapter 3 also illustrate the search for supervenient properties. Think of Gause's *Paramecium* experiments. Whether a given *Paramecium* survived or not depended both on the medium into which it was introduced and on the presence or absence of other protozoans. Any given *Paramecium* cannot be simply described as feeding on certain bacteria: for whether or not it does so depends on other factors, like the existence of competing protozoa. Again, the existence of *guilds* poses interesting problems about the supervenience of ecological traits. A guild is a group of species which seem to exploit the same environmental resources in similar ways. At first sight, the existence of such guilds would seem to vindicate the claims that ecologically significant characteristics are indeed supervenient. For if choice of prey is determined not just by genetic programming, instinct and anatomic structure but also by the offerings of the environment and the presence of competitors, then we would expect the partitioning of resources by a guild to be a consequence of perhaps subtle differences among the relevant species and their competitive interactions. In terms of the theory of the niche, we might expect these differences to show up in the *realised niches* occupied by the different guild species.

However, research on guilds has not always delivered the straightforward results that we might expect. It can be very difficult to determine whether the various species in a guild are currently engaging in, or in the past have engaged in, any form of competition. Additionally, there is not always a direct correlation between behavioural or morphological features and the nature of the realised niche. However, some studies of guilds do seem to deliver results congenial to the claim that ecological roles supervene on the other characteristics described. A study of African tree squirrels, for example, showed that the niches occupied by nine relatively similar species could be plausibly differentiated by the use of four dimensions — the habitat type, the height at which foraging occurred (in the canopy or at ground level), the nature of preferred foods and, finally, the body sizes of the squirrels themselves (see references to Emmon's work in

Begon, Harper and Townsend 1986: 679–80). The squirrel's place in an ecological community thus seems to supervene on a relatively small number of characteristics of its own structure and genetic program and on features of its environment.

As hinted already, the situation is more complicated when we turn to modular organisms, like trees and shrubs. Although these items have a program that determines their growth, the program is highly sensitive to features of the environment in which the organism occurs. Of course, this is just what we should expect: a sessile organism must adapt to its environment or die, while a motile one has some choice about which habitat to adopt. One surprising result for modular organisms is that some aspects of their morphology are not among their basic properties. For example, a white clover plant shows differential patterns of growth depending on whether or not it is growing in a grassy area (Begon, Harper and Townsend 1986: Ch. 1.5.1). The lengths and branching characteristics of clover stolons are thus supervenient properties of the plant, varying with environmental factors, rather than basic characteristics.

We can now start to see why there are different accounts possible of what a plant or animal is. In terms of species membership, we can describe plants and animals as belonging to this or that species. To do so is to say what *kind* of thing the plant or animal in question is. In traditional terms, the *natural kind* of an organism is an *essential*, not a contingent or *accidental* feature of it. Hence, I belong to the kind *human*, and could not be the kind of thing I am if I were not human. We could put the point in terms of possible situations. Since my humanity is essential to me, then I could find myself in situations in which I am of a different weight, or a different height from the weight and height I currently am; but there is no possible situation in which I could find myself, or imagine myself, where I am not of the kind *homo sapiens*.

Of course, things have many non-essential properties of significance. It is not essential to me that I write, or am a parent, although I regard both of these features to be of very great significance in my life. For organisms in general, morphological features are the basis of species classification, and thus constitute their most fundamental essential properties. To be a member of the species *Solidago* (golden rod) a plant must have the sort of stem, leaf, flower and root structures that golden rods display — and, more minutely, have one specific kind of DNA resident in its cells. An elegant study of golden rod populations by Abrahamson and

Gadgil showed that those present on open, dry sites allocated a far larger proportion of biomass to flowers than those in damp woody sites (Putman and Wratten 1984: 272–4). It makes sense that open dry sites, which are prone to (and often the result of) disturbance should contain populations making a higher investment in reproduction than do those in less disturbed sites deep in woodland. Drawing parameters from the logistic equation (see Chapter 4.2) we can distinguish populations of r-strategists (high-volume reproducers) from those of K-strategists (low-volume reproducers). Species and populations may be classified according to whether they are r-selected (small, high adult mortality, short generation time, tracking vicissitudes of the environment) or K-selected (large body size, long lifespan, small reproductive allocation, steady population). In these terms, what is interesting about golden rod populations is their pursuit of different strategies in different conditions.

There are many problems with the r/K models (Begon, Harper and Townsend 1986: Ch. 14.9–14.11). Independent of the general applicability of the idea of distinct reproduction strategies, it seems significant that different populations of the very same species adopt quite distinct strategies for the different environments in which they are found. It follows that a certain degree of investment in reproduction is not an essential feature of members of the species *Solidago*. Reproductive tendencies of individual plants and populations of them are thus context-dependent, just as being a writer or a parent are features of mine that are context-dependent. Think now of the difference between realised and fundamental niche. If the distinction has any significance at all, then some organisms will differ in their realised niches according to the situations in which they find themselves. No realised niche, then, can be an essential property of the organism that occupies it. If there are essential properties to be found in connection with niches, then it will be the fundamental niche that is necessary for defining an organism in ecological terms — that is, the fundamental niche may be something the ecologist counts as an essential characteristic of a species and its members.

There are difficult problems lurking here. The discussion that has just taken place might make some readers dubious about the whole distinction between essential and accidental properties. In taking such a stance, they would be siding with some influential modern philosophers and looking askance at what Quine calls the 'metaphysical jungle of Aristotelian essentialism'. On the other hand, our practice of classifying organisms by species or by ecological role (or in any non-arbitrary, scientific way) suggests that there are principles of classifying and modes of explaining, which are in some way non-accidental. The point being laboured here is that, however we choose to describe our procedure, the ecologist is able to contribute to our understanding of what something is. But this contribution will involve reference to the context, the environment and the interactions which are significant for this mode of explanation.

In one way, it simply does not matter whether we call facts about ecological role essential or merely contingent (or accidental). The important point is that what something is and does depends in part on where it is. A description of an organism's significant properties, dispositions and life-strategies will involve more than an account of its morphology, genetic structure and other species-relative properties. This is not to diminish the importance of the latter sort of property. But there are other modes of describing what kind of thing we are studying and of explaining its behaviour. Is it an organism that feeds on insects or on plants, and is it one that will maintain a viable population given the introduction of a population of a different species competing for the same resources? To these questions, we give answers not in terms of what species the organism belongs to but in terms of its *ecological role*. And, as we have seen, this role supervenes on characteristics that are internal to the organism and on others that are external to it. Populations belonging to different species can play the same roles in a certain sort of community, while populations belonging to the same species may well play different roles in different communities. Community roles, ecological functions, or whatever we want to call these characteristics, are thus important supervenient, rather than intrinsic, properties of organisms.

It may seem, though, that there is a disanalogy between the supervenience of ecological traits and the original example of supervenience — that of colour on surface material and texture. The disanalogy may appear all the greater when we consider other favoured examples from the philosophy of supervenience. Consider, for example, the supervenience of the moral on the natural. In terms of this notion, two persons who are alike in all their natural properties cannot be different in terms of their moral properties — one cannot be good while the other is bad. Or think of the supposed supervenience of the mental on the physical. Although different physical states may support exactly the same mental state, a change in mental state cannot occur (so it is said) without a corresponding change in physical state. These notions of supervenience may seem at first to be radically different from the biological one. For it may seem, in the moral case, that it is logically or conceptually necessary that moral properties supervene on natural ones — not a matter of empirical discovery, but more a matter of what it means to be good or decent. And in the second case, that of the mental and the physical, it seems that, as Kim puts it, the relation between base and supervenient properties is stable across possible worlds (Kim 1984). But ecology deals in empirical, not logical or conceptual matters; and predator–prey relations and other significant ecological traits are capable, as we have seen, of being different (that is, there will be possible situations in which, with the same base properties instantiated in the organism, different supervenient ones will be present).

To go into these issues in suitable detail would require close study of various kinds of supervenience relation which would take us far from our topic here (but see Blackburn 1985; Kim 1984; and McFetridge 1985). Whatever the relation between traits of organisms and environment in virtue of which the organisms have their ecological roles, the term 'supervenience' is fine for describing it. Moreover, the problem about accident and essence is a general one in science, not just specific to ecology. There is something non-arbitrary about the laws of mechanics, the principles of thermodynamics and the discoveries of optics. Yet the principles encountered in all these branches of science could be denied without any logical contradiction. We can label the puzzle here by distinguishing logical from physical impossibility, but such labels take us no closer to a grasp of what kind of necessity we are identifying in our theories about nature. If there were not some empirically discoverable but non-accidental relation between the basic properties of object and environment and the kinds of characteristic of interest to the ecologist, it is doubtful if ecology could claim to be a science at all. For any science must aim to elucidate connections among phenomena, connections that are not in any sense arbitrary, or random, but which are constitutive — at least in part — of what it is to be a negative ion, a freely accelerating body or a large predator. The 'design stance', mentioned in the last section, could only be adopted — even as a pretence — given the supervenience of ecological role on more

118

basic properties of organism and environment. If this conclusion looks reductionist to some, then so much the worse for the anti-reductionist. Understood correctly, however, this section has revealed what truth there is in holism.

8.3 Ecological communities

It is a commonplace of ecological folklore that biological communities and ecosystems are equilibrium aggregates of diverse species living in complex interrelationship with each other. However, a glance at scientific ecological findings, and knowledge of current debates in species biogeography, reveal that real situations seldom live up to the myth. Worse, the description of real situations is hampered by unclarity over such central notions as 'diversity', 'community', 'stability', 'equilibrium' and 'ecosystem'. In this section, I will try to shed what light I can on these often puzzling matters.

When Gause grew species of *Paramecium* in test tubes he was bringing into existence, we might say, new *experimental entities*. What this means is that such test-tube systems had not existed before and are certainly not found in nature. There was a vogue in philosophy of science until quite recently of wondering about the whole status of *theoretical entities*. These are things whose existence is postulated or mentioned in theories, and whose behaviour is meant to explain observations and data which, without their existence, are not so well explained. Such entities are also of the sort that have not been hitherto observed, or which are unobservable given the available technology. Even though tubes containing *Paramecium* might not have existed before ecologists became interested in competitive exclusion, there is little merit in challenging their claim to reality. However, when theories mentioned electrons, genes, positrons, or other 'small' items for the first time, there may seem to have been a genuine question about their reality. The problem about theoretical entities thus came to be one about the real existence of unobservable, or hard-to-observe, items (but see Hacking 1983 for a challenge to conventional wisdom).

Biological communities and ecosystems seem to have a status that is not quite as clear as that of experimental entities, nor so doubtful as that of theoretical ones. A forest, pond, or grassy field can be easily enough identified, treated as separable from its surroundings and considered as a unit for community studies. The problem is not so much one of reality but of usefulness. If community ecology is to say anything of interest about such systems of populations, then there has to be some way in which one such system can be segregated from others and described as having properties of biological substance. As Horn's results have shown, there is a real puzzle to be raised not about the reality of forests, but about their possession of characteristics which contribute to our understanding of the distribution and abundance of individuals and populations within them. Shift the focus and think of a small patch of earth round one tree in a forest: the roots, microscopic fungi and soil nutrients are again only worth identifying as a system if such a way of thinking produces theoretical benefits. The problem of the theoretical status of communities and ecosystems is therefore a real one, even if there is no problem about reality.

We can think of communities and systems in either a 'top down' or 'bottom up' fashion. Suppose, to take the latter mode first, we try to specify what makes species population P_1 a member of some same community as population P_2. Using the notion of supervenience from the last section, we can define the relation between P_1 and P_2 as follows: P_1 is a member of some same community as P_2 when some supervenient property of P_1 depends on the existence of P_2, or vice versa. For example, a certain species population can only be a population of *grazers* given another population for them to graze upon (likewise for other predator–prey relations, like parasitism and for commensalist and symbiotic relations). Biologists, who do not operate with the notion of supervenience, sometimes make a similar point by talking about the *functional* classification of species populations.

The definition of co-membership just given is partially, but not viciously, circular. Ecologically supervenient properties are, after all, precisely those that depend on environmental interactions. Despite its circularity, the scheme is helpful in drawing attention to two further points. The first is that the same scheme applies equally well to relations involving a species population and abiotic items: P_1 belongs to the same ecosystem as abiotic items a_1, a_2, . . ., a_n when some supervenient property of P_1 depends on the existence of a_1, a_2, . . . a_n, and vice versa. It might appear that the 'vice versa' is inappropriate here; but abiotic items — water, carbon, nitrogen and so on — do have supervenient properties (e.g. of cycling in determinate ways) that require the existence of the various populations which support the cycling. Second, according to these schemes, the sun is an element in every ecosystem. Notice that the definition does not commit us to looking for aggregates where every population interacts with every other. Every population in a community or ecosystem interacts with at least one other population in the same community. This allows that communities may be fairly loosely bonded, that a population can belong to many different communities, and that some communities can straddle or sprawl across others. Precisely this kind of flexibility is required in order to make sense of the rough and ready way ecologists split up the world into communities. No definition, perhaps, would be expected to do real justice to the ability of the trained ecologist to group populations in interesting ways according to the aspects of community or system of life being investigated.

The existence of a scheme involving the abiotic that is similar to the one involving only the living perhaps explains the characteristic emphasis that some ecologists have put upon *ecosystems* as the fundamental ecological unit. Modern emphasis upon *communities* is not so much a change of direction in ecology as a simple shift in emphasis. Whether we focus on the ecosystem or the community, our proposed definitions leave open an empirical question of some significance. The question is whether we will find in nature any clear cases of communities or systems that are relatively self-contained — as aggregates of items whose supervenient properties are mutually fixed in an interesting way — or whether there is no sensible segregation to be made short of the entire global ecosystem itself. I do not know a definitive and unchallengeable answer to this question, but I will proceed as if there are in nature suitably self-contained systems and communities whose segregation from others can be divined by the ecologist.

However, this bottom up approach to community structure leaves out many commonly recognised aggregates. A taxonomic group within a habitat might attract attention (for example, all the phytophagous insects in a stretch of moorland). This group would still be called a 'community' even if there is little mutual fixing of supervenient properties. Likewise, any ecosystem of reasonable richness will contain populations which have no direct ecological effect on each other, and may not even interact with the same abiotic resources. The 'top down' approach will count such assemblages as communities if there are patterns of organisation to be found that are common to numbers of such assemblages. In other words, the community investigator looks for non-arbitrary patterns that are characteristic of community types. The scientific move is, as always, to consider the particular as an instance of the general. At a global level, deserts, moist tropical forests, temperate grasslands, coniferous forests and the like are all identifiable types of community (or *biome*). These massive kinds of community are perhaps the ones for which patterns of *succession*, *productivity* and *diversity* seem most naturally sought. Between these massive communities which are partly based on geographic divisions and the type of community best thought of in terms of mutual interaction lie a whole host of intermediate cases.

From the 'top down' perspective, it makes sense to ask questions like the following: are there systems and communities which show similar structural features, even though consisting of radically different species populations? Are there correlations in general between diversity and stability? Are there generalisations concerning productivity and consumption that hold for very different kinds of community? Are communities always structured according to competition, and do isolated or island communities show different structures from less isolated ones? Are there general principles governing differentiation of niches within systems? In fact, all these questions have been tackled and answers proposed by community ecologists. If communities and systems are to figure in ecology as anything more than aggregates inheriting their properties from the properties of their components (see Chapter 6) these, and other such questions, must have biologically interesting answers. In some cases — for instance, the problems concerning the composition of island communities — the issues are obscured by the very real difficulties about formulating appropriate null hypotheses (see Chapter 3.4). In other cases, interesting generalisations seem possible (Begon, Harper and Townsend 1986: Chs. 16–18).

To conclude the present section, let us reflect on one problem posed by community studies. Popular, or 'folk' ecology, has it that nature manifests a balance or equilibrium sustained by diversity within communities. The notion of equilibrium here is not the familiar one found in thermodynamics; for in thermodynamic terms, natural communities are far from equilibrium and maintain their non-equilibrium status due to continual fresh energy inputs. Nor is it clear that systems and communities are stable, if by this is meant that they maintain some fixed ratio of species populations or productivity. What ecologists and systems theorists mean by *stability*, however, is the tendency of systems to return to some previous state after perturbations (Emlen 1984: 61ff). Any system that persists through time may manifest varying degrees of stability, fluctuation and oscillation. In fact, *resilience* rather than stability may be a more important notion, where by 'resilience' is meant the 'ability to absorb change and disturbance and still maintain the same relationships between populations or state variables' (Holling 1976: 81). Such resilience may involve many complex interactions among populations together with flexible behaviour on the part of their members. For example, a general principle of predation might be to eat a broad diet when food is scarce, specialise when it is abundant and never forego a food of higher energy value in favour of one of lower value (Emlen 1984: 444). Populations operating according to this and kindred rules may be able to contribute to the resilience of the community within which they are located.

However, it is not clear that diversity itself contributes either to resilience or stability. Think of competitive exclusion or of the forests studied by Horn. Were it not for fire, storm or other disturbance, the most successful competitors might be thought likely to dominate a community, driving the other populations to low — or even zero — density. But many real natural communities and systems, even those undisturbed by humans, show enormous richness and diversity of species (where *diversity* involves a reasonably even distribution of individuals in a community among the species populations). Clements argued, as we have seen, that the process of succession has only a small number of outcomes. Mathematical models of the Lotka-Volterra variety apparently confirm this view, at least as long as we consider communities as relatively closed systems. Real communities, of course, suffer disturbance at various times from a variety of natural causes. It is sobering to reflect, however, that there may be in many communities and natural systems an underlying tendency towards stability or resilience at the expense of complexity or diversity. If we value diversity or complexity in communities and ecosystems this may not be because such features are natural. If these are good things, we can strive to preserve them, by interfering, if necessary, in systems that have a natural tendency to decline in diversity (Begon, Harper and Townsend 1986: Ch. 21.2).

8.4 Community role and the Gaia hypothesis

This chapter has so far introduced two important ideas. The first, and perhaps the more basic, is the idea that the traditional account of an item's essential nature leaves out many of its significant properties. These depend on its behaviour towards and relationships with other things in its environment. We have called such significant properties 'supervenient' and these can be thought of in terms of community role. The complete biological story about what an item is thus includes its species membership (essential properties) and also its community roles (supervenient properties). Part of the story of what an animal or plant is will be an account of where it is (and hence of what it does) — an account making essential reference to its biotic and abiotic surroundings.

We thus give an ecological twist to Shoemaker's theory of properties as clusters of conditional powers (see Chapter 6.2; and Shoemaker 1984: Essay 10). One of his examples concerns the property of being *knife-shaped*. In combination with other properties (like the properties of being made of steel, knife-sized, and so on) being knife-shaped is the power to cut wood, slice butter, and so on. Now an object can clearly have such properties in a world that happens to lack wood or butter. What it cannot have in such a world is the chance to exemplify the first-order power to produce changes in bits of wood, portions of butter and other things that knives regularly modify. If the knife does, however, cut butter as part of its usual functions, then typically cutting butter would be an example of what I have called a supervenient property.

The second idea is that there are facts about communities, patterns to community structure and ways of describing biological systems in general which reveal communities to be more than inheritors of properties of their components. If this second thought is along the right lines, then biological communities, and hence ecosystems, are structured, persisting items whose behaviour is to some extent predictable. To say this, of course, is not to deny that we can explain the patterns and behaviours of communities and systems in terms of their component populations and chemistry (see Chapter 6). A further implication of the second idea is that we can link it in interesting ways to the first. If communities constrain, and are in turn constrained by, the features of their members, then the account of where an item is not only tells us about that item but also throws light on the nature of the system or community in which it is placed.

Real communities are often very complex. We can, however, simplify some of these considerations and see what they mean by restricting our attention to communities of just a few members. Some of the most interesting of the simple cases involve a phenomenon I have so far discussed hardly at all: *symbiosis*. Whereas the study of populations in macroscopic communities is often associated with models of competition, it is not entirely misleading to think of microscopic entities in terms of mutualistic models. Mutualism and symbiosis often escape attention simply because the items involved are so fine. What we think of as the roots of plants are, in many cases, not roots at all but *mycorrhizae*, an intimate interweaving of fungus and plant tissue. The mycorrhizal fungi gain carbon in return for making available mineral resources to the host (Begon, Harper and Townsend 1986: Ch. 13.7). However fine the discriminations necessary to uncover such mutualism, the associations themselves are of vital significance in the maintenance of higher plant and animal life.

As well as mycorrhizae of plants, it is well known that the digestion of cellulose, xylose and starch in the guts of ruminants depends on microbial fermentation. There are whole communities of specialist gut bacteria and protozoa whose interactions are of considerable ecological interest. But if we look more closely at the micro-organisms to be found in the intestines of various creatures, we find further mutualistic associations of a striking kind. In the intestines of termites and cockroaches, for example, cellulose is broken down mainly by specialist protozoans. These protozoans, when first observed, were thought to be fringed with fine hair or lashes — *cilia*. This is not surprising, given how they look (Fig. 8.1). In fact, these fine hairs are now known to be *spirochaetes*, bacteria which live in close association with the protozoans. The mutualism here is striking. The motility of the protozoan depends on the spirochaetes; in return for providing propulsion the spirochaetes receive nutrients from the protozoans.

It is a well known principle of evolutionary theory that new items are in some instances constructed from *stable sub-assemblies*, that is items enjoying a separate and independent existence. The association of protozoa with spirochaetes in the termite gut provides a clear example of such a process in action. Lynn Margulis has recently advocated that a similar process accounts for the evolution of *eukaryotic cells* (cells with nuclei). Such cells are the building blocks of all animal and plant life and any degree of complexity. Cells without proper nuclei — so-called *prokaryotic*

Figure 8.1: Symbiotic protozoa in termites

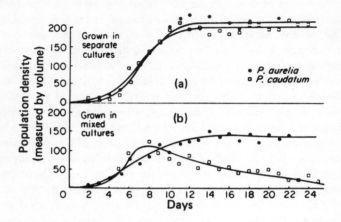

Symbiotic protozoa from the intestinal tract of termites. These drawings were made at a time when these organisms were thought to be ciliates. The cilia are now recognised as spirochaetes living in a mutualistic association with their protozoan host.

Source: Begon, Harper and Townsend 1986

ones found in microbes, bacteria and blue-green algae) — do not go in for constructing many kinds of multi-cellular organisms. By contrast, nearly all the familiar plants, moulds, animals and micro-organisms (like protozoa) are built from nucleated cells. 'Sexual' cell division — involving contributions from more than one 'parent' in producing offspring — is a feature of eukaryotic cells (Fig. 8.2; and Margulis 1981: Ch. 2).

Cells with a nucleus display a complex architecture. What is the origin of this architecture, which features a nucleus in its own membrane, various endoplasmic *reticula* (which can transport proteins to the exterior of the cell), *mitochondria* (containing enzymes for respiration), *plastids* (responsible for photosynthesis in plant cells), and — often — mechanisms of motility like *cilia* or tails? According to Margulis, the organelles of cells are the result of symbiosis: 'Eukaryotic cells are considered to have originated as communities of co-operating entities that had joined together in a definite order; with time, the members of the co-operative, already skilled in their specialities, become organelles' (Margulis 1981: 3).

Although this particular theory of cell evolution was not

Figure 8.2: Comparison of prokaryotic and eukaryotic cells – composite diagrams based on observations at the light and electron microscopic levels

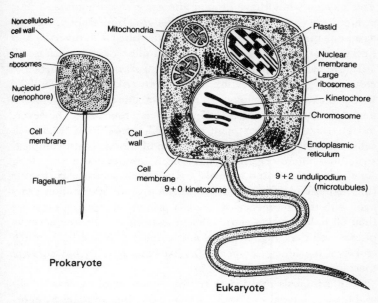

Source: Margulis 1981

originated by Margulis, she has defended it in a number of places (Margulis 1981 being a clear exposition of the theory). If correct, the theory is not only extremely interesting in its own right, but also suggestive of how important mutualism and symbiosis may be to the development and maintenance of life on earth. The history of prokaryotic cells is ancient; it goes back more than three billion years to the cyanobacteria discovered as fossils in limestones in Zimbabwe (Margulis 1981: Ch. 5). By around 700 million years ago, Margulis argues, the main trends in cell evolution had occurred, involving symbiosis among the precursors of cell organelles. During that time the carbon, nitrogen and hydrogen needs of living things would have been met at first by the ammonia, carbon dioxide and methane available as atmospheric gases. Before oxygen appeared as a significant atmospheric gas, cell life was steadily evolving, and the bacteria present were already carrying out impressive functions, including fermentation, anaerobic photosynthesis, nitrogen and carbon-dioxide fixing, and oxidation of hydrogen sulphide to sulphur.

About two billion years ago, oxygen started to accumulate in the atmosphere, oxygen produced by biological photosynthesis. Free oxygen would have limited atmospheric ammonia and thus reduced the 'greenhouse' effect of this gas. Importantly, oxygen as a toxin put selective pressure on prokaryotic life to diversify and find ways of coping with its effects. Bioluminescence, which is still a feature of many living things, may be a strange consequence of certain oxygen-detoxification mechanisms (Margulis 1981: 129). Under the influence of oxygen, new metabolic pathways occurred and also the first eukaryotic cells. Only very much later, however, did any kind of higher plant or animal, constructed from such cells, emerge. In the overall timescale of life on earth complex, structured micro-organisms have been around for a very long time, while plants and animals are relatively recent experiments (the 'explosion' of higher forms began around 500 million years ago, with the emergence of modern primates occurring no more than 60 million years ago). Looked at from the time-perspective of cell evolution, higher plants and animals may not be so much 'survival machines' for genes (as Dawkins puts it) but survival machines for cellular life.

The symbiosis that led to the development of nucleated cells and the subsequent explosion of higher life thus predates by enormous amounts of time the symbiosis among complex developed things that are commonly cited in the textbooks. Among these the mutualism of fungus and alga shows the same feature of emergence that is characteristic of the eukaryotic cell. The lichen has characteristics and features that result from the interaction of fungus and alga; they are not inherited from properties of either on its own. As Margulis puts it, 'Lichens are remarkable examples of innovation emerging from partnership: they possess many morphological, chemical and physiological attributes that are absent from either partner given independently' (Margulis 1981: 167). The presence of such emergent properties can make the presence of symbiosis at the microscopic level hard to detect. The ciliate *Paramecium*, *P. aurelia*, one of the subjects of Gause's experiments described in Chapter 3.3, was thought for some time to have a genetic feature (labelled *kappa*) which failed to obey Mendelian principles. After much study, the kappa particles were found to be symbiotic rod-bacteria and not part of the genetic material of the protozoan (Margulis 1981: Ch. 7).

Although we possess mathematical models for competition, there are few useful models for symbiosis (Begon, Harper and

Townsend: Ch. 13.12). This is unfortunate. The major chemical cycles of the plant and the balance of atmospheric gases are, as far as is known, sustained critically by micro-organisms. Prokaryotic bacteria and other microbes produce and control the reactive gases of our atmosphere, an atmosphere that supports life only because its proportions are maintained within very fine tolerances. So far from competition being the ultimate determinant of life on earth, it seems that symbiotic associations are the *sine qua non* of any life support system. Life requires co-operation for its continuance.

It is a short step from these reflections, some of them still speculative, to the formulation of the *Gaia hypothesis*. In brief, this is the theory that, so far from life occurring when environmental conditions were favourable, life itself is responsible for producing and maintaining those conditions that support life. The salinity of the sea, and the oxygen content of the atmosphere, to take but two examples, have been constant for billions of years, and this very constancy requires explanation. Suppose atmospheric oxygen levels increased, for example, to 25 per cent (four per cent above its standard value). In such a case, the next lightning strike, or match struck, could initiate a fire that would consume material of very high moisture content (plants in tropical forests, for example). Little of our planet's vegetation could survive. Lovelock, the originator of the Gaia hypothesis, uses examples like this to suggest that life on earth is a complex system, governed by elaborate feedback mechanisms, which thereby ensures its own continuation (Lovelock 1979).

If the Gaia hypothesis were true, the entire biosphere that is Gaia herself could be thought of as one large, self-regulating organism, with redundancy, feedback channels and spare capacity enabling it to maintain life in some form in the face of numerous contingencies. Even catastrophes on the scale of those which eliminated the dinosaurs and two thirds of the species living at the time were not enough to destroy Gaia's life-preserving potential. Global thermonuclear war, though disastrous for humans, other higher animals and many plants, would likewise be something that Gaia might well survive.

It is not clear to what extent we can regard Gaia as possessing 'emergent' properties, that is, properties that are not simply inherited from her parts. In terms of the account developed in Chapter 6, there is nothing anti-reductionist in the idea of emergent properties, for these need not be novel or unpredictable. What is puzzling, however, is the extent to which Gaia is enough like a lichen, a nucleated cell or other conceivably unitary item to count as one herself. We have to be wary of thinking from the design-stance about the various ways in which earth seems anomalous when compared with the other planets. The stability of gas concentration in the atmosphere and other such features are not clear evidence of cybernetic systems in operation. To borrow an example from Andrew Woodfield, consider a tank of water whose contents are heated by some source and which stay at a constant temperature. Although such a result could be obtained by the presence of a thermostat or other feedback device, it may not be. It might just be that heat gained by the water from the heat source exactly balances that lost through conduction (Woodfield 1976: Ch. 11).

Another worry arises if we subject some of Lovelock's recent work on the hypothesis to scrutiny. By modelling the temperature control achievable by a number of daisy species in a 'daisy world', Lovelock has recently claimed to have come up with evidence that species diversity in a cybernetic system gives very fine degrees of control (Watson and Lovelock 1983). If true, this is an important result. However, there is a much weaker thesis that could also be maintained on much the same evidential base as used by Lovelock. Instead of regarding Gaia as a kind of superorganism, we can use the information about earth's enormous deviation from the pattern of other planets' atmospheres, geology and hydrospheres as evidence for the claim that the maintenance of higher life forms depends on the continued existence and flourishing of certain 'lower' life forms. Having mentioned the importance of kelp in the production of iodine, a substance vital to healthy mammalian life, of algae in the production of sulphur and of the anaerobic bacteria of wetlands in producing oxygen, Lovelock writes:

> The really critical areas which need careful watching are more likely to be the tropics and the seas close to the continental shores. It is in these regions, where few do watch, that harmful practices may be pursued to the point of no-return before their dangers are recognised; and so it is from these regions that unpleasant surprises are most likely to emerge. Here man may sap the vitality of Gaia by reducing productivity and by deleting key species in her life-support system; and he may then exacerbate the situation by releasing into the air or the sea abnormal quantities of

compounds which are potentially dangerous on a global scale. (Lovelock 1979: 121)

We do not need to imagine Gaia to be a unified superorganism in order to appreciate the importance of regulating the salinity of the sea, the oxygen content of the atmosphere, the volume of the ozone layer, and so on. If we can identify the natural mechanisms which control these, and other life-supporting functions, then we would do well to leave them well alone. If we are unsure of the impact of our mining, farming and consuming practices on these mechanisms, we would do well to move cautiously. To destroy a number of estuarine mudflats may be to do more than wipe out a few billion colonies of prokaryotes. It may be to do damage to a planetary life-support mechanism.

What I have just suggested is at odds with the idea that the life-support mechanisms of the planet are of necessity complex and densely interrelated. Instead, human, higher plant, and other animal life may be itself an adaptation to environmental equilibrium that is maintained by a relatively small number of systems. Suppose, as has been suggested, that the temperature of the earth is importantly regulated by oceanic cloud cover and this, in turn, is regulated by the amount of solar energy reaching certain algae inhabiting the oceanic surface. Here we have the ingredients for a feedback mechanism that may be of some significance in setting the parameters for life on earth. If our fishery and waste-dumping practices threaten the algae in question, then the disruption of this one critical mechanism would not wipe out life on earth but might well alter selectional pressures in new and dramatic ways. If we like things the way they are, we would be well advised to leave such critical systems alone.

Whether there are simple systems of such critical importance is something that is hard to determine. It may be that work like Prigogine's on self-regulating, or dissipative, structures would be of relevance to Lovelock's hypothesis by showing how a complex enough system can maintain non-equilibrium, but stable, conditions, even in the face of random, Markovian changes in its subsystems (Prigogine 1976, 1980). My suspicion is that such investigations would show that, from the design-stance we can think of the biosphere as an entity with at least one property in its own right, namely the resilience to maintain itself in states that are far from chemical or thermodynamic equilibrium. However, to try to defend or evaluate such a claim lies well beyond the scope of a philosopher's professional competence.

In the last three chapters, we have come to important conclusions regarding ecology, the properties of organisms and the kind of perspective that scientific ecology can provide and which could conceivably make an impact on our ethical and political thinking. Perhaps the most important discovery has been the extended account of the natures of things that follows from taking ecology seriously. The biologically rich story of what something is tells us what it typically does in various situations. Significant properties will thus depend on where an item is nested in its community or ecosystem. The framework of ecology enables us to make sense of the 'functional' properties of living things in terms of how they fit into larger systems or communities. The Gaia hypothesis provides yet another input to our modes of thinking about humans and their place in nature. We have existed for only a minute amount of time compared with the history of other high forms of plant and animal life. Yet we know that we are able to interfere in powerful ways with global systems, including those of significance in maintaining Gaia herself (however unitary or otherwise she happens to be). In the following chapters I will be arguing that, however rich or thin an account of human nature we give, we should not ignore the perspective on this supplied by scientific ecology.

9
Theory, Fact and Value

9.1 Science and ethics

There are obviously several ways in which ethics may impinge on science. For example, certain areas of research may raise ethical problems for the investigator or would be investigator. Is the suffering to the animals involved justified by the importance of a specific piece of research involving cats? Is the risk of nuclear devastation made more or less likely by certain kinds of weapons research? Given that a significant portion of the world's population is underfed, badly housed and lacking in decent education and health care, can we justify allocating enormous sums to work in high technology industries whose products are only of use to those who are already indulged, overfed and unregenerate consumers of global resources? These questions and a whole host of kindred ones, are familiar to everyone who has ever given a thought to the state of the planet and its inhabitants. To embark upon an industrial, financial, technological or research career without asking them would be the first step in leading an unexamined life, a life that does less than justice to the humanity of the person living it.

What I wish to consider in this section is the converse influence, the influence that science might have on ethics. This will involve enlarging on the sketchy remarks made earlier about sociobiology's contribution to our perspective on human nature. By 'ethics', I mean not just the principles that we claim to follow in action, but the ideals, kinds of intentional behaviour and roles that we regard as characteristic of a person's moral style. For the moment I do not want to make firm distinctions between the ethical on the one side, and the political and aesthetic on the other, nor even between the ethical and the moral (as does Williams 1985). When the scope of what is ethical is so vaguely characterised, there are likely to be many different influences upon it. Although we might class all such influences as cultural, one way or another, it is useful to try to separate out the scientific from the others. We can thus ask whether the biological sciences in particular provide a framework within which certain distinctive ethical attitudes, styles of response and principles of conduct are motivated.

To suggest that the sciences may make a significant impact on ethics (whatever the culture) is to disagree with those who regard ethics as having a peculiarly intimate connection with the religious and philosophical tradition in a society. Thus Passmore, for instance, in his book *Man's Responsibility for Nature* argues, in effect, that any environmental ethic that is to take root in western countries must be based on features of the Judaeo-Christian tradition. As he puts it at one point:

> The fact that the west has never been wholly committed to the view that man has no responsibility whatsoever for the maintenance and preservation of the world around him is important just because it means that there are 'seeds' in the western tradition which the reformer can hope to bring to full flower. (Passmore 1974: 40)

Passmore's original work in this area has been criticised by several other writers. In the case of Robin Attfield, at least part of the debate between him and Passmore concerns the degree to which the Judaeo-Christian tradition is one which motivates arrogant dominance or humble stewardship on our part towards nature. Having discussed the nature of a metaphysics suited to our contemporary environmental concerns, Attfield writes:

> . . . the Judaeo-Christian tradition . . . is not essentially productive of metaphysical arrogance, despite the fact that such arrogance has often besmirched it. Nor is it committed to regarding man as 'apart from nature'. Indeed, it embodies indispensable insights about human capacities and obligations, such as that people are the custodians and stewards of a precious natural order, and have a creative role in actively enhancing it. (Attfield 1983: 63)

One immediate difficulty about this particular debate is the problem of grasping just what is at issue. It is not clear why certain Biblical sources should loom large in the discussion of influences on our contemporary ethics, any more than the writing of the Stoics or of medieval theologians. In fact, any thorough historical analysis of attitudes to nature (for example, Thomas 1983) will look at the impact of theology, the life sciences, the physical sciences, artists, poets, moralists, philosophers, farming practice and agricultural technology, patterns of trade, the relative role of different social classes, and a whole host of other data. I see no reason why what applies to understanding our attitudes to nature, should not also apply to understanding our ethical attitude, if any, towards nature.

What kind of role, then, can be suggested for the scientific study of ecology in forming ethical attitudes? For simplicity's sake, let us think of a case where a practical decision has to be taken in keeping with some general principle. Not all moral behaviour is of this deliberative kind. But it will clarify some of the issues of concern in the present chapter if we focus on this kind of case. In coming to a decision on such a matter, we have to have two kinds of information available. On the one hand, we must grasp the rights and wrongs, or the principles, involved in a situation where action is required. On the other, there is the need to know relevant details about the nature of the case — the persons or other animals affected, the relationships among them, and so on. In crude terms, we could distinguish between the *factual* and the *evaluative* aspects of such situations.

If we stick, for the moment, to this simplified picture, then we can notice a clear role for scientific findings. Our evaluations, planning and behaviour will be more sophisticated the more we know about the relevant facts. Our knowledge of the impact of our decisions and actions may well make a difference to the weighting we give to the items before us in our deliberations — a weighting that might well have been different before we knew all the facts. For brevity, let us call this effect the *extension of awareness*.

The extension of awareness made possible by the sciences can involve more than simply a clearer knowledge of the consequences of our actions and a better grasp of the means whereby we can achieve various ends. Think of the impact of Freud's theory of our unconscious desires and motivations. What his work suggests is a quite precise way of delimiting the scope of human freedom, and of defeating the claim that someone is responsible for a particular action. We come to see that some behaviours are compulsive or obsessional, and not to be assessed ethically as on a par with normal behaviours. Freud's theories go further: they threaten our conception of *altruism*, and his psychological determinism can easily generate a pessimism about the prospects for any very great freedom, sincerity or spontaneity in human action. Without going as far as this, we can see the possibility for a reinterpretation of existing ethical categories under the impact of some scientific findings. This possibility of revision is in many ways more significant than other forms of extension of awareness, and we will look into it in some detail.

As we saw in the first chapter, sociobiology is also capable of supporting a *reinterpretation of the moral*. As with Freudian psychology, its impact is *subversive* in that if we take it seriously we have to reconsider the whole position of 'pure' morality. By 'pure morality', a notion I borrow from Williams, I mean an account of the moral life that emphasises the freedom, dignity and rationality of moral agents. It is precisely those features that the sociobiological, and the Freudian, stories subvert. What I now wish to suggest is that ecology is capable of funding significant extensions of awareness — including a revision of values — in a different direction. It is a science that offers the prospects of extending our awareness in ways relevant to practical decision-making, and at the same time, it has the potential for reinterpreting some of the categories that are of fundamental importance in our ethical thinking.

9.2 Reinterpreting the moral

In what way, then, can the biological, or life, sciences extend our awareness and make a revisionary impact on our moral beliefs and attitudes? In answering this question, we have to be careful not to expect too much from any particular framework of thought — or any scheme drawn from a set of frameworks. What I mean is that we cannot expect to be able to deduce new moral beliefs, attitudes or ideas directly from such frameworks. Rather, they provide a setting, a background, within which certain attitudes make sense, and from which they derive support. Ecology is in a particularly powerful position since it is a fine example of an interdisciplinary subject. It thus provides us with an image of the world which is drawn jointly from the resources of biology, geology, chemistry and biochemistry.

From biology, the ecologist is able to derive an understanding of physiology, reproduction and the life histories of various plants and animals. In fact, all of these areas intersect with areas of direct ecological concern. For example, physiological changes in a raptor due to the assimilation of toxic substances from its typical prey may lead to it becoming less fit in an evolutionary sense. That is, its ability to reproduce may become impaired, or its offspring may themselves be too damaged to survive long enough to pass on their genetic material. The impact of the toxic material in the food web containing this raptor population is thus likely to lead to an ecologically significant change — namely a new distribution of species populations in the community or ecosystem under study.

The combined biological and ecological information about the simple situation just described now funds a further claim that can be made. Clearly, the raptors in the imagined system are not flourishing, and therefore the toxic material in the food web is having a deleterious effect on them. We have not yet given any strict biological or ecological definition of terms like 'good', 'bad' or 'deleterious'. Let us say that what is good for an individual hawk are — roughly — those conditions which allow it to follow the life cycle and activities typical of its kind; what is good for a population of hawks is harder to specify, but in rough terms we could think of it as the conditions in which the members of the population all have an opportunity to flourish.

We can think of the matter in *teleological* terms, that is in terms of organisms having goals that they seek. Paul Taylor puts the point in this way:

> We conceive of the organism as a teleological centre of life, striving to preserve itself and realize its good in its own unique way. To say it is a teleological centre of life is to say that its internal functioning as well as its external activities are all goal-oriented, having the constant tendency to maintain the organism's existence through time and to enable it successfully to perform those biological operations whereby it reproduces its kind and continually adapts to changing environmental events and conditions. (Taylor 1986: 121–2)

Taylor here builds in the notion of 'good' in terms of an organism's realisation of its biological program. Whatever account we eventually settle on as being apposite for the analysis of teleological explanation, it is clear that the notion of goal-directedness is at home in the biological sciences. The frustration of an organism's goals is likewise naturally regarded as bad for it — though of course to say such a thing is not to suggest that such frustration is bad in any ethical sense.

The intimate knowledge of organisms in their natural state, and understanding of their physiology, their role in biological communities, their capacity for flourishing — their ability, in short, to pursue what Taylor calls their 'good' — is a way of providing a context within which an attitude of care about natural things makes sense. It does not follow that every creature or organism that has a good, is good. We are a long way from recognising respect for the anopheles mosquito or supporting the Aids virus in its struggle for survival. But we are in a position to make sense of extending our moral concerns beyond the merely human, to make sense of a serious degree of respect for other living things, and — what goes with that — a recognition of the need to limit our activities where their freedom to pursue their ends is threatened. An especially serious threat of this sort is one that may destroy a population of individual organisms, while the ultimate threat is one leading to the extinction of a genetic lineage.

We are still far short of articulating anything like Leopold's land ethic. The considerations in this section do no more than suggest the possibility of funding an extension and reinterpretation of morality by appeal to the life sciences. But it is already clear that the sorts of extension of awareness now mooted are in sharp contrast to the kinds of case previously discussed. The impact of sociobiology or of Marxism seemed likely to diminish our humanity, assimilating our behaviour to types found among less free, less rational creatures. If we are no more than products of biological or socio-economic evolution then we are less than we normally like to think we are. If, as suggested in this section, other living things have goal-oriented modes of existence, have modes of flourishing or suffering, conditions of freedom or restraint which are to some extent like ours, then we may be motivated to think that they are more than may be implied by our usual treatment of them.

However, the perspective described in this section, the *biocentric* one, to give it a name, is not one with which we can ultimately rest content. It asks both too little, and too much, of us. Moreover, it involves notions about the worth of natural things that require close attention. In the rest of this chapter, then, I will subject this view to some strenuous criticism. The criticism in turn forces us to think about the grounds for judgements of value, merit or worth, and raises the question of just what sorts of things are possessors of these features. These concerns lead naturally on to thinking about ethical theory in general. Let us make a start by looking in the next section at the issue of value as it arises on the view of the 'deep' ecologist.

9.3 Deep ecology and ethics

More than one chapter could be dedicated to unravelling the core doctrines of the so-called 'deep' ecology movement. Like many others, I have been struck by certain aspects of the deep ecology position, and sympathise with some of its goals. As I hope to show, the position as a whole suffers from an attempt to give simplistic answers to complex questions, as well as lacking an overall unity and coherence. This is not a problem unique to this particular movement. Exactly the same criticism will be made of contemporary ethical theory, though for different reasons.

As far as we know, humans are the only kind of creature to adopt moral postures, be motivated by moral concerns and to act with due regard to the needs, interests, cares and responsibilities of other agents. Human action, then, has the potential for a richness denied the behaviour of other known creatures. As I, and others, have pointed out elsewhere, it is not a consequence of this point about human agency that humans themselves are the only proper objects of moral attitudes (some would say 'subjects' with the same intention — namely to refer to those items upon which moral attitudes of care, consideration, sympathy, and the like are focused). (For an account of what might be called the *moral considerability* of non-human beings see Goodpaster 1978; Attfield 1983; and Taylor 1986.)

What is the nature of moral standing? It is the *value* that something has by virtue of the fact that concern for it enters, in a certain constraining way, into the deliberations of a moral agent. The way in which the item figures in the agent's deliberations has to be spelled out carefully. If I conceive it as my duty to visit an old friend in difficulties, one constraint on my deliberations about how to make the visit may be the information I have about rail services between where I am now and where my friend is. But this does not show that rail services have any kind of moral value. We can, however, start from the easy cases, and work outwards. When another agent figures in my moral deliberations, I will be concerned with things like that person's interests, needs, responsibilities, responses, duties and so on. What constrains my choice of behaviours involving that other agent in the ideal, simplified case will, specifically, be my conception of that person's interests. The fact that other agents do have interests, just like me, has led some thinkers (for example, Feinberg 1974) to claim that only items with interests can be possessors of rights, and thus be represented as suffering benefits and harms as a result of my behaviour. I would deny that anything like the point about rights holds for moral considerability or standing. For just as constraints on my choices of action may be imposed by my recognition of interests on the part of some other person, this recognition motivating a concern on my part that these interests are protected, so there will be constraints on my choices of action motivated by concerns I have about things that have no interests, or whose interests could not be easily divined.

It might be objected that genuine concern for the continued existence, or maintenance, of an item is no evidence that such an item is morally considerable, or a possessor of any kind of moral value. But suppose my concern shows all the seriousness that is sometimes taken as the hallmark of the moral: my deliberations and actions are constrained in an especially compelling way by concern that the item in question be preserved, be left undisturbed or be maintained in a certain condition. Argument is now needed to show that my concern here is not moral, that the item is not receiving moral consideration from me and that the item does not therefore possess moral value. Suppose it is asked whether I am willing to *universalise* my deliberations and their constraints. I agree that any agent in a similar position to me ought to be constrained by a concern in respect of the item in question of the sort I have. Again, we are faced with clear evidence that the item figures as morally considerable in my thinking. Finally, suppose it is asked whether my concern with the item is bound up with a concern to maximise the utility, happiness, freedom from pain or welfare of humans, or of humans together with other living things. This time, the question requires a more sophisticated response, but only by way of distinguishing instrumental from non-instrumental value. If the item in question, however, is one that I deem to have *non-instrumental* value, then my concern for it will be unrelated to my concern for the utilitarian issues just mentioned. In considering the possibility of such concern, we get the first hint of the irrelevance of utilitarian thinking to some significant moral issues.

Against the background sketched above we can now characterise the position of the deep ecologist as one which develops the central theme that things other than humans, or humans and a select group of other animals, have value or worth of a non-instrumental kind. Some deep ecologists would be suspicious of this characterisation of their position, this being based on two considerations. First is the fact that in Arne Naess's original article (Naess 1973), which spawned the cult's title, there is no mention of this specific claim about value, although there is a manifesto which specifies deep ecology as committed, among other things, to egalitarianism in the biosphere (everything has an equal right to live and flourish), to a total field holism (organisms are knots in the web of life), to rejection of class divisions in society, to rejection of pollution and resource depletion, and to the encouragement and maintenance of diversity and complexity in natural systems. Second, there is a general question over whether the core of the deep ecologist's position should be taken as involving a valuational position (as argued by Sylvan 1985) or as involving a metaphysics which happens to make a certain value position plausible (as argued by Fox 1984).

However interesting these issues are for students of the deep ecology scene, they are conveniently irrelevant to my concerns here. The metaphysical position of the deep ecologist has already been surveyed and found wanting. As we saw in Chapter 4, Capra espouses an idealism which requires specifically philosophical defence, and certainly does not flow from quantum theory in any of its standard interpretations. In Chapter 5, we saw that issues of reduction and holism are generally of less significance than they are often thought to be, and since the metaphysics of the deep position are regularly described as anti-reductionist and holistic, it will be clear that they do not hold much prospect of giving a decent grounding to claims about natural value.

For the purposes of this discussion, I will take the content of the deep position as given in two ways. First, we can use the following sets of contrasting slogans (based on Naess 1984):

Shallow ecology	*Deep ecology*
Natural diversity is valuable as a resource for us	Natural diversity has its own (intrinsic) value

It is nonsense to talk about value except as value for mankind	Equating value with value for humans reveals a racial prejudice
Plant species should be saved because of their value as genetic reserves for human agriculture and medicine	Plant species should be saved because of their intrinsic value
Pollution should be decreased if it threatens economic growth	Decrease of pollution has priority over economic growth
Third World population growth threatens ecological equilibrium	World population at the present level threatens ecosystems, the major threat being posed by the population and behaviour of industrial states more than by those of any others. Human population is today excessive.
'Resource' means resource for humans	'Resource' means resource for living beings
People will not tolerate a broad decrease in their standard of living	People should not tolerate a broad decrease in the quality of life but should be ready to accept a reduction in the standard of living in overdeveloped countries.
Nature is cruel and necessarily so	Man is cruel but not necessarily so

Second, here is a list of further deep slogans which Naess supplies later in the same paper:

Animals have value in themselves, not only as resources for humans
Animals have a right to live even if of no use to humans
We have no right to destroy the natural features of this planet

142

Nature does not belong to man
Nature is worth defending, whatever the fate of humans
A wilderness area has value independent of whether humans
have access to it

Although Naess takes these statements as being specially germane
to a certain meta-ethical doctrine — objectivity of value — this
particular aspect of his position does not concern me here. The
collected statements are meant to indicate in a rough and ready
way the kind of platform occupied by the deep ecologist. Despite
the criticisms I will make, I will later be reconstructing a view
which does justice, I hope, to many other serious ethical ideals
associated with the Naess position.

There is one version of deep ecology, however, whose moves
are of interest because of the problems they reveal in allocating
value to natural things in their own rights. In some of his recent
writings, Naess has suggested *self-realisation* as the core idea of the
deep position. This fits nicely with the strong holistic claim that
I am merely a knot in a biospheric web, a momentary formation
of energy in an energy field. As such, I interpenetrate other
biological objects and processes, exchange energy with them, and
am constantly changing in response to changes in my environment. Self-realisation, as Golley has put it, reflects the insights
and experience of the field ecologist but expresses them in a
personal idiom (Golley 1987).

The value correlate of all this is that as I come to know and
appreciate what I am, so do I come to know and appreciate the
natural things around me with which I am in close association.
Rather as Callicott put it (see Chapter 3) the self comes to be
extended to embrace the things around it; not all my parts are
under my skin. But if I am valuable, then so too is the system
within which I exist, for I am — in some mysterious way — one
with it. However much metaphysical nonsense may be thought to
be involved in all this, there is certainly one good thing to be said
for such a metaphysics. It does overcome the old problem of how
to find value outside the valuing subject. For by building in the
items of the subject's environment, by breaking down the divide
between the self and the other, we have a simple solution to the
problem of value in nature. Provided I am valuable, then so is my
extended self, the natural world. A wrong to it is a wrong to me.

But notice the obvious weakness in this move. We have
succeeded in finding a solution to the problem of natural value

only by opting for an *anthropocentric* perspective. The initial problem was whether we could really maintain that things other than moral agents were themselves items of moral standing. The reinterpretation of the moral did not, however, involve extending the scope of morality to include things other than the moral agent. On the contrary, all that this move achieves is an extension of the moral *agent* to include the rocky crests, the forests, the mountains and rivers, and all the other items we count as figuring in our moral concerns. The resulting morality is no improvement on anthropocentric morality: the radical extension of the agent is not a genuine extension of moral considerability.

A similar criticism can be levelled at another aspect of the deep platform. On some accounts, its core value assumption is the claim of biospheric impartiality or egalitarianism. This view is motivated by Naess's original claim that all living things have an equal right to live and flourish. In some recent treatments, this view has been sensibly modified so as to permit the elimination of hazardous bacteria and viruses, and even of larger organisms that threaten the existence and flourishing of other organisms. However, we can see that what is probably intended by such a principle is an end to human *chauvinism* (although see Williams 1985: 118 for observations on the way such a label can mislead).

Without going into detail about how to formulate a principle of biospheric impartiality that would yield moderately plausible results, and leaving other criticisms of the biocentric viewpoint to the following chapter, let me consider a more general worry. Not only is the extended conception of self hospitable to *egoism*, the principle of biospheric impartiality threatens to reinstate such egoist notions at one remove. For instead of taking the individual agent as the focus of interest, the biospheric impartiality principle takes the welfare or benefit of *living* things, as of special moral significance. But by so doing, it still restricts the class of morally considerable items to those that have life and that together constitute the web of life whose existence we have to protect. Williams puts a similar point: '. . . the requirements of benevolence or fairness may always stake a claim against self-interest; *we* can represent a self-interest as much as *I*' (Williams 1985: 15). In these terms, the issue becomes one concerning the range of the expression 'we'.

This latest, and very general, point indicates something extremely important about morality. My suspicion is that, in the end, all moralities will involve *identification* on the part of the agent with items or groups other than the agent. This will be a topic for later exploration. For the moment, notice that we have started to characterise problems that arise for the deep ecological perspective. Even if we ignore the peculiar metaphysics, we are faced with problems about the interpretation and plausibility of that version of it which involves a biocentric approach. The limitations of this whole approach deserve separate treatment and to this task I now turn.

10

Puzzles about Value

10.1 Biocentrism and worth

The biocentric perspective has been persuasively urged by Paul Taylor in his recent book *Respect for Nature*. His discussion promises to provide some useful vocabulary in terms of which we can explore one set of ideas about value and worth in nature. As we will see these ideas are contrary to the spirit of some major themes in contemporary moral philosophy — a finding with implications of some importance. Important though the ideas are, Taylor's own distinctions between value, worth and merit prove inadequate to funding the ecological dimension to ethics.

Although I have already talked, loosely, about the value of things, distinguishing instrumental from non-instrumental values, it might be objected that, speaking strictly, all value is in the mind, or at least the experience, of the valuer. We would be moved to accept a view like this if we agreed that without valuing agents, there would be no values. Suppose that value is something we project on the world, rather than something there to be found. Then, again speaking strictly, there can be no values in objects that we value: rather, we allocate value to things, we are the source of the value, and — we might well add — what is primarily valuable is our own situation or experience in respect to the things valued.

We can characterise one account of value along these lines by using distinctions suggested some time ago by C.I. Lewis. He wrote:

All value in objective existents is extrinsic; it consists in a potentiality of the thing for conducing to realization of some

146

positive value-quality in experience . . . Intrinsic value, which is that for the sake of which all other things are valued, belongs exclusively to occasions of experience as such; and value in objects consists in their potentiality for contributing goodness to such occasions. (Lewis 1946: 432–3)

Lewis's account gives us a degree of clarity. An experience which is valuable in its own right is one possessing what he calls 'intrinsic' value; the object giving rise to the experience can be either 'inherently' valuable — when the intrinsically valuable experience is realised simply through presentation with the object — or instrumentally valuable, when the item in question simply leads to the presentation of a further, inherently valuable, object. For the sake of expounding Taylor's and Lewis's conception, the terminology just introduced will be used in this section. Thereafter, for reasons that will become clear, much looser terminology will be employed.

A problem now arises about items that may have no role as primary content in valuable experiences but which are of a sort that we might nonetheless deem valuable. Another arises in connection with the restriction of intrinsic value to experience — could we not count actions, projects and even situations of certain kinds as being intrinsically valuable? The second problem is relatively minor. Let us suppose, for the sake of the argument, that interests, actions, pursuits, as well as experiences, can all be counted among the things that can have intrinsic value. Of course, many of our activities and pursuits have no such value: we do them from habit, or as a means to some other end. But it makes sense to think of at least some of our activities, projects and experiences as being of value in and of themselves.

The first problem demands further thought. Lewis himself makes the distinction between the instrumentally and the inherently valuable. But what of those items which do not figure in our lives under either guise but which are nonetheless things with a good of their own? Taylor suggests that, instead of distinguishing different kinds of value, we use the term 'inherent worth' to capture this idea of things having their own good (Taylor 1986: 75). As we have already seen, such items are precisely those for which we can define better or worse conditions of existence, and for which we can recognise the difference between freedom to flourish and constraints on their natural modes of developing and reproducing. All living things, in Taylor's view, will be items of inherent worth.

Following a proposal due to Gregory Vlastos, Taylor then distinguishes *inherent worth* from *merit*. Human beings, and other living things, may have various merits in varying degrees. One shrub may have the merit of screening the fence behind it just as someone may have the merit of being a good dentist. But merit is independent of inherent worth. All the items that we class as 'persons' have the same inherent worth, according to Taylor, even though they differ widely in their merits. To be possessors of inherent worth is to carry a special status, namely that it is wrong to be treated merely as means to someone's ends. Taylor then argues that the recognition of inherent worth in every living thing carries the same moral burdens (Taylor 1986: 77–9). It is wrong, then, to treat any living thing merely as a means to someone's end. There are other moral constraints here as well. In particular, he argues that the promotion of the good of any living thing is itself an ultimate end, to be brought about simply for the sake of the being whose good it is; and, further, it is a matter of moral principle that moral agents give consideration to the good of beings that have inherent worth (in other words, this consideration should not be dependent simply on love of, or inclination towards helping, the being in question).

At this stage, I want simply to indicate the presence of certain very general problems with Taylor's biocentric approach which limit its appeal in quite significant ways. On what basis, for example, is the claim to equality of worth for all living things to be founded? It is true that all living things show a tendency to maintain their existence in the face of environmental pressures, although such maintenance varies from the relatively simple, to the sophisticated homeostasis found in higher animals. Many communities of living things likewise seem to show the ability to maintain community existence by being resilient in the face of disruptive pressures. But a community of living things is not itself a living thing and hence is inherently worthless by Taylor's criteria. How then am I to weigh considerations regarding my family against considerations regarding trees in my garden, or the river along which I walk most mornings? For the trees each have inherent worth, but the river, like my family, seemingly falls into the category of the worthless.

If Taylor's theory distributes worth in the wrong way with respect to families and rivers, then it follows that his account will be silent about issues that are of serious moment when we think about the things that count morally for us. I will not repeat here arguments that I have used elsewhere (Brennan 1984) to show that there is no principled means of regarding the class of living things as the class containing all and only the things that have a clear claim to moral consideration. To forestall a misunderstanding, it is important to recognise the difference between concern we might have regarding some item; and duties, obligations and respect that we may owe directly to the thing in question. I may be concerned, for example, not to drive into my neighbour's gate; but any obligation or duty involved in this concern is owed to my neighbour and not to the gate. My duty is one simply *regarding* the gate. In seeking to characterise natural worth or value, I am interested in concerns that are more than object-regarding. Rather, the concerns I have in mind involve, if not duties and obligations, at least consideration and respect owed directly *to* natural things, whether living or not.

10.2 Systemic value

The terminology and distinctions suggested by Lewis do not cast light on the topic of value in nature. Even if we could establish that some natural objects play essential parts in activities and experiences that are intrinsically valuable in the sense defined by Lewis, this would at most provide grounds for respect or duties regarding them, not for any respect owed directly to them. As long as experiences are taken as primarily valuable, the value of the objects of such experiences is bound to be secondary. No special value need attach to a person or work of art in virtue of causing an intrinsically valuable experience. At least in principle, a replica or other substitute could bring about an exactly similar, valuable experience (E. Katz 1985). Our lack of success forces us back to consideration of the claims on behalf of *worth*. Perhaps we should concentrate our attention on giving an account of worth which is relevant to, and motivates, moral concern for items that have worth. So far, this option has been dismissed with the observation that the claim that an item has worth, in Taylor's sense, does not of itself have any moral implications.

If we now return to vocabulary that is somewhat loose and vague (like most useful vocabulary!), we can consider Rolston's attempts to fund a perception of moral worth or value (Rolston

149

1986, 1987). Although he distinguishes instrumental from intrinsic value, Rolston uses the terms 'value' and 'worth' as synonyms, and makes no attempt to root the valuable primarily in experience or activity. On the contrary, he tries to draw on a conception of value funded by ecology:

> Never have humans known so much about and valued so little in the great chain of being . . . We rationalize that the place we inhabit has no normative structures, and that we can do what we please . . . What if, in truth, we are not only limited by the natural facts but also by natural values? What if living well is not merely a getting of what I value, but a negotiating of my values in a neighbourhood of worth? (Rolston 1986: 114–15)

Notice how well these remarks fit with some of the things deep ecologists represent. Naess, as already noted, has made suggestions about the objectivity of value, and the issue of subjectivity versus objectivity in value has exercised Rolston for some time (see the essays in Rolston 1986). In psychology, it is common to distinguish the subjective (or *concept-driven*) contribution to knowledge from the objective (or *data-driven*) side of the story. A similar distinction seems natural in respect of moral worth, value or standing. To what extent, we wonder, is the special status of the morally considerable item derived from its properties rather than imposed on it by our own cares and interests? The answer I will propose involves questioning the very question just put. There is no subjective–objective dispute here because both sides of the story are required. Value, or moral standing, is a phenomenon that arises in a dynamic context of interactions. Ecological insight will enable us to transcend the either–or implicit in the question without forcing us into an implausible metaphysics.

Rolston, more than most philosophers in this area, writes with a first-hand understanding of natural processes and with an awareness similar to the ecologist's. He recognises that all living things are the product of historically particular forces, that chance, rather than design, has shaped the evolution of populations, and that things are what they are by being located in natural surroundings. Remember the argument in Chapter 8 that ecological properties are supervenient. Rolston, although not expressing himself in this vocabulary, is aware of the important truth that such a claim is meant to express. He is also sympathetic to the general thrust of Leopold's land ethic. The result is an attempt to develop a notion of *systemic value* that will fund the appropriate attitude of respect for nature. Aldo Leopold, in a much quoted passage, writes:

> . . . quit thinking about decent land use as solely an economic problem. Examine each question in terms of what is ethically and aesthetically right, as well as what is economically expedient. A thing is right when it tends to preserve the integrity, stability and beauty of the biotic community. It is wrong when it tends otherwise. (Leopold 1949: 224–5)

We can in fact distinguish two different theses that might be attributed to Leopold. One is that the biotic community, or 'the land' as he sometimes says, is an item of value in its own right. The other is the less interesting thesis that simply recognises the immense instrumental value of the land. Let us take the Leopold thesis, however, as the stronger one.

Rolston sets out to show that ecosystems are valuable in their own right, and we can think of his remarks as applying equally to biological communities as well. In order to argue that ecosystems have more than merely instrumental value he suggests we need a new category of value — systemic value — which attaches to systems and is distinct from the value associated with individual things. We can follow his thinking on this, and start to understand his claim that ecosystems are responsible for *heightening* individuality by considering three features he identifies in ecosystems.

1. Ecosystems are wild in that they are decentred, open, loosely organised, and beset by historic contingencies. The historical particularity of each biological community is logically and empirically entwined with the individuality of each inhabitant.

2. Evolutionary ecosystems have, over geological time, steadily increased the number of species on earth to over five million.

3. Not only has the quantity of individuals and species increased over time, but so has the quality of life. In

particular, evolution has produced mobile, warm-blooded, neurally complex creatures, and creatures with sentience and self-consciousness. Each stage of evolution has been a liberating development, a freeing of individuals. This liberty allows individuals to move from one ecosystem to another. (Rolston 1987)

We could summarise these features by saying that the ecosystem *places*, *produces* and *promotes* individual things. Notice that the claims about heightened individuality parallel those already mentioned about the enriching or locating effect of the system on the individual.

Now suppose that we take the centred complex life, situation and experiences of beings such as ourselves or the other higher animals as being of value in its own right. Ecosystems promote the new arrival of just such life. We might say that ecosystems select for just such heightened individuality. This would give us good grounds for supporting a weak instrumental thesis. But, Rolston seems to argue, it also gives us grounds for supporting a stronger thesis. If we are not to mistake the fruit for the whole plant, then we must not limit our duties simply to ourselves and other centres of consciousness, sentience, or whatever. We have to recognise value in the system itself.

Before coming to grips with this large claim, we have to recognise some limitations in Rolston's three points. It seems perfectly possible to imagine a world with fewer organisms and less diversity of species than ours contains, but which would be a better world. Such a world might contain, for example, no slums, either human or natural. Also, if high quality self-conscious life is a really good thing, then perhaps massive reductions in other species could be tolerated to make room for more of us — a thought that is bound to be disturbing to any preservationist. If his third point about quality of life is taken seriously, we might also wonder if Rolston has really broken free from human chauvinism. If it is because of high quality products like ourselves that the system is given its special value, then is he not simply declaring the system valuable because we are?

10.3 Two modes of valuing

We operate with two different modes of valuation, one of which locates value in individuals, the other of which locates it outside individuals. It is because we do have these two modes available that Rolston is able to suggest grounds for projecting value onto systems. Let us think of the individual knots in the web of a system as beings which carry their own share of value. Rolston writes:

> . . . the community history is not merely that of its species, although the history is written there. The history is not all privately in individuals . . . History is smeared out across the system. That which is coded in the DNA sequences of the birch tree, or that of the individual coyote's career, is clever, but that which is diffused in the biotic community as a matrix of co-evolved historical centres is equally remarkable . . . The context and impetus for history is as much the system place as it is the member inhabitants. (Rolston 1987)

Let us translate these considerations, for a time, into social ones. There is a way in which I carry, like everyone else, a share of the culture, ideals, aims and history of my society, my family, the institution in which I work, and so on. I therefore carry marks, so to speak of various groups and communities. Now, since I live in a liberal society, different life plans and different projects are allowed to be equally good. So some of the projects, ideals, standards, enthusiasms, loves and hates that I carry, keep alive and occasionally pass on to others will be regarded as worthwhile. On this view, then, each of us embodies some community values — doing things, and caring about things which are important in, and fostered by the communities in which we live.

Note that there is an ambiguity here in the notion of community value. This could be either of two things. First, we could be referring to situations, activities and projects which any serious-minded person in the community would acknowledge as worthwhile in themselves. These would be community values in the sense of securing wide assent in the community. But, second, we might be indicating projects and enterprises of the community as a whole, which individuals may participate in, but which the community possesses in its own right. It is this second sort of community value which I will take Rolston as trying to characterise.

In terms of the community values we each embody, we can say that I have a different mix from each of you. Nonetheless I can recognise good and valuable things in the lives of others. In terms reminiscent of Plato's *Euthyphro*, my life might be thought good to the extent it embodies projects and ideals that are good; but my projects and ideals are not made good simply because *I* am pursuing them. This is an important point. It is undeniable, however, that we also adopt the view that projects and activities acquire their value from the individual engaging in them. This means that we are committed to the idea that individuals have value in their own right. Think, for instance, of our feelings when we lose someone near or dear to us. Part of what we have lost is the unique blend of projects, ideals, characteristics and so on that were distinctive of that person — though no single project, or other feature, was unique to the person. But we also notice the loss of things that may have had no value in themselves, but gained value from association with the person in question. A typical smile, a figure of speech, a certain mannerism — these are things that might have been irritating, or have gone unnoticed, when displayed by others but which have acquired value through association with the person in question.

Now let us try to reintroduce some Leopold-like considerations in the light of the two ways of valuing people. Think, first of all, of the identification of our own interests with the interests of a wider community or group (Naess 1984). As I have already suggested, we may not be able in the end to give any less self-interested an account of value than this. To understand more of what is meant by *identifying* ourselves with other things, let us suppose something like the following thesis to be true. Where I identify my interests with those of some wider group (whether family, political party, university department, professional body, or whatever) there will be a perspective on which my activities are — or could be — instrumental in promoting the group's own good (whatever that is). This is too strong as it stands, but let us leave the qualifications till later. The issue for the Leopold thesis is then this. When we think of larger biological communities, or systems, is there a perspective we can adopt on which we see ourselves as being instrumental in promoting the good of the system? Leopold's recommendation that we promote the beauty, integrity and stability of the biotic community may be thought of as a recommendation for us to identify with a community that extends far beyond our social, or species groups.

Leopold's extension of ethics amounts on this reading to the view that we can identify with the land itself — with ecosystems,

conceived in Rolston's way as loose but structured collocations of
the living and the inanimate. Were the Gaia hypothesis true, this
identification might be easier. But we do not need to view
ecosystems as organism- or person-like in order to be able to identify our interests with their value. The burden of much writing in
environmental ethics has been to provoke such identification.
Reading Leopold, or reading Rolston, we are drawn to identify
our own good with that of the land; we feel ourselves enriched by
such an identification, just as we are enriched by finding ourselves
identified with other groups.

If I am right in reading this kind of position out of Leopold
and Rolston, then we can make sense of the relevance of scientific
ecology to the defence of such a position. It gives us an account,
even if not a definition, of the fundamental notions of *ecosystem*
and *community*. If I am what I am partly by being in a certain
environment, part of a certain community, then scientific ecology
can spell out at least some aspects of the value I carry and the
value I donate. For, as well as my relation to my society and to
various social groups, I am also part of several biological
communities.

Do we then have grounds for optimism about Rolston's defence
of systemic value? Can I identify with the natural systems around
me so as to consider ways of promoting their good or harming
them? Unfortunately, the answer I fear has to be negative. A
natural system, like a natural community, has no good of its own,
and so cannot be harmed or helped in its own right. Some moral
philosophers have, as already noted, taken it as obvious that only
items with interests have moral standing, for they can be
represented (either in person or by others acting on their behalf)
and they can be benefited or harmed by our own behaviour
towards them. Taylor's move, of making the *worth* of an item
depend simply on its having a good of its own went some way
towards defusing the force of this kind of argument. For we need
not say that a simple micro-organism has interests, provided we
recognise that it has modes of growth, feeding and reproduction
which can be affected by our interference. But we cannot even
make sense of such an idea in application to communities and
ecosytems.

We have already observed that biological systems and
communities are sometimes a matter of chance: what enables a
system to persist a while may well be the fact that it is no more
than an adventitious collection of organisms that are able to live

together and exploit the available resources. Should new, immigrant populations appear and drive out the original members of the system, or should environmental conditions change (perhaps through resource depletion or climatic disturbance) a new organisation will probably come to replace the old one. This is the inevitable process of change that characterises the development of life. But how, in the face of such considerations, can we make sense of the idea of the good of the system or community itself? Social communities may have a good, for there are better and worse ways for them to develop. But natural communities have no ends to serve, no purpose in their development, and no goods of their own. Of course, we could always deem the good of a system to be that kind of development *we* want to see. This is an entirely straightforward way of thinking about the good of a natural system — but such goods would be the aesthetic and other goods of our own, not goods of the systems themselves. The attempt to fund moral respect for nature on some notion of systemic good or value thus has to be abandoned.

11

The Environment and Conventional
Moral Theory

11.1 The tragedy of the commons

Chapter 9 began by considering the impact that the sciences might have on ethics. To start this one, I want to look at one biologist's attempts to argue for an extension to ethics by appeal to facts about carrying capacity. Garrett Hardin's essay, 'The Tragedy of the Commons' (reprinted in Hardin 1972), has, since its original publication in 1968, been one of the most widely cited works in discussions of the economic, political and ethical dimensions of environmental issues. The original essay aimed to show that coercive policies may be required to keep human population at an optimum level — the optimum being well below that at which a maximum human population can be sustained. Hardin's general argument is that coercive policies, mutually agreed within a community, are the only ones that will protect common resources. His theory is of special interest in the context of the comparison of teleological with Kantian moral theories that I am about to undertake. Although I argue that neither account gives us what is required by an environmental ethic, the contractarian or Kantian position comes closer to giving a plausible account of our ethical situation. The truth of this claim is nicely shown by considering Hardin's work.

Hardin's example concerns a mythical, uncontrolled common on which a number of herders keep cattle. In the light of the freedom of access he imagines, it might have been more historically accurate for him to have written about the tragedy of the range, since common pastures in medieval England, for example, were not uncontrolled in the way required by the story. However, suppose that herders have free, unlimited access to the pastures in question. As long as the number of cattle using the common stay below the carrying capacity of the range, there are no problems. According to their individual wealth, the several herders can add additional cattle, gaining the benefit of fattening their herds on the free range. But now imagine that one day the carrying capacity is reached — that is, that the total herd on the common is the maximum that can be sustained without overgrazing and erosion.

Each herder is motivated by rational self-interest, let us suppose. Moreover, as the carrying capacity of the range is exceeded, the initial damage is slight — leading, for example, to the cattle on the range not putting on quite as much weight as before. The benefit to each herder of adding an extra beast to the common is far greater than the disbenefit of having marginally thinner cattle. For the benefit accrues to the individual herder, while the effects of overgrazing are shared by all who use the common. As Hardin puts it:

> Adding together the component partial utilities, the rational herdsman concludes that the only sensible course for him to pursue is to add another animal to his herd. And another . . . But this is the conclusion reached by each and every rational herdsman sharing a commons. Therein is the tragedy. Each man is locked into a system that compels him to increase his herd without limit — in a world that is limited. Ruin is the destination toward which all men rush, each pursuing his own best interest in a society that believes in the freedom of the commons. Freedom in a community brings ruin to all. (Hardin 1972: 254)

It hardly needs this story of natural disaster to support the claim made in the last sentence. Unbridled freedom is not an option in any community, although it is interesting to inquire into the origin of constraints on freedom and the source from which coercive power derives its authority. On the Kantian view of morality it is the rationality of individuals assenting to life in a shared community which is the ultimate source of the constraints that such community life involves. Put in terms of the imagined herders, the Kantian or *contractarian* point would be that forming a community with other herders involves accepting certain norms governing the behaviour of everyone in the community. If one of the norms prohibits actions that threaten the carrying capacity of the range, then an individual who breaks that norm, but accepts the benefits of life in a community governed by such norms must expect, and accept, the appropriate penalties.

Although the contractarian theorist can thus describe the situation in a way that starts to make sense of each individual's responsibility for protecting the common resource, the *utilitarian* faces a graver difficulty. On utilitarian grounds, actions are right when they maximise happiness, or utility, over a population. We can think of the choices facing an individual herder in terms of the following dilemma that is fatal for any simple version of utilitarianism. Suppose our herders are all well aware of the limits of the range they are using. Individual herders thus have an interest, we might think, in keeping their own herds at a reasonable level, provided others do likewise. But either the other herders are going to maintain sustainable populations of cattle, or not. Suppose they do. Then a smart herder can reason as follows: if my fellow-herders keep their herds at a reasonable size, I can add a few cattle to my own herd, thus maximising my own interest, without that action causing excessive disutility all round. In fact, the common losses might be negligible.

Any herder who reasons in this way and accepts the overall principles of utilitarianism would have to conclude that the right thing to do is to increase the numbers of cattle on the range. But what is to stop other herders thinking along similar lines? If they do they will fail to maintain their herds at a sustainable size, and then our original herder has to ask the following question: what point is there in my keeping my own herd within reasonable limits when everyone else is going for the short-term profit? It is hard to give any utilitarian reason for limiting one's own use of the resource when it is being plundered by others.

It follows that, if I am a utilitarian herder, I should go all out to maximise my short-term interest. For either my own actions lead to a negligible deterioration in the common good (first option) or I am a sucker for restraining my own use of the resource that is being overconsumed by my neighbours (second option). These considerations add force to the arguments used by Hardin. For Hardin's own case involves simply a group of people who are motivated by rational self-interest. Our development of his case, however, involves a herder committed to what is ostensibly a *moral* position of some substance. In the face of the dilemma, we have a number of choices. We can stick with the version of utilitarianism, and accept that respectable moral positions may be ecologically disastrous (and disastrous in any other case where the sharing of common resources is concerned). Alternatively, we can give up the position completely, or try to modify it in ways that retain the credibility of the underlying position. An attempt at the latter was J.J.C. Smart's famous recourse to game theory (1973) in order to deal with this kind of problem. Simply exploring it will, I hope, reveal the moral emptiness of such versions of utilitarianism.

The 'tragedy of the commons' problem can arise wherever it is desirable to limit use of, access to, or pollution of, a shared resource. The common in question can be outer space, national parks and nature reserves, clean air, rivers, common fisheries and so on. Should I conserve water during a drought, when my prize runner-beans are at risk? The local authority, let us suppose, has put a ban on the use of hosepipes. Others will be conserving water, or they will not. Either way, the sort of reasoning just explored would lead me to conclude that my continued use of a hose will not, on its own, make the drought worse; so I should continue to water my beans, thus increasing my own, and the aggregate, utility. But if everyone does the same, there will be no water for anyone. Smart suggests that, since it is true that some people can continue to use water as normal without posing a problem for supplies in general, what a society of dedicated utilitarians might do is give each person a certain probability of not obeying the hosepipe ban (Smart and Williams 1973: 57–62. It is only fair to point out that Smart recognises that such a strategy is likely to be purely theoretical, and in real cases 'good' utilitarians would simply obey the ban. His case, incidentally, is one concerning energy, rather than water.). I then throw dice, or use a random number generator, or other technique by which I decide to break the ban only in the event of some highly improbable outcome being realised. But, if everyone else does the same, the result will be that most people in the society will observe the hosepipe ban, and only a few break it — thus achieving the desired result.

What is so bizarre about this solution to the difficulty is the idea that we have here a characterisation of an ethical course of action. There could be circumstances in which throwing dice constitutes a perfectly proper, and ethical, decision procedure. But if I am a member of a community and share certain common concerns with the other members of it, then I am likely to want to do my bit to help that community when it faces difficulties. Later, I will be arguing that our membership of our society and of the natural communities in which we live is not all that it might be. But if membership means anything at all, it has to involve the possibility of doing things *as a member of the society or group in question.* Those who take seriously the prospect of throwing dice to determine whether or not to help the community in times of difficulty are ignoring the whole dimension of membership, participation and involvement that — as I will be arguing — is central to our very existence as moral beings.

We could think of the same case in a slightly different way. Suppose that I try to get away with disobeying the hosepipe ban, but get found out. There is a fine for breaking the ban, which I am now obliged to pay. Have I grounds for resentment or complaint? It seems to me that the simple utilitarian does have grounds for complaint, in the light of the dilemma already posed. But if I am in the position of the rational, assenting agent so beloved of social-contract theorists, my situation is quite different. My assent to live in, and reap the benefits of, society involves assent to accepting the penalty which has just been imposed. For no society would be just in which those who were known to have gained advantages of this sort over others were able to escape unscathed. In assenting to life in a just society, I also assent to the necessary degree of coercion to ensure that just rules are obeyed. The rule about conserving water was precisely of this sort — one obedience to which is to the benefit of all. By contrast with this kind of view, the utilitarian one seems designed almost to subvert the very attitudes, feelings and dispositions upon which social cohesion is founded. It is a morality for those who have no conception of society as anything beyond a mere aggregate of individuals.

Unfortunately, it is a long way from these kinds of considerations to ones that reflect, with any degree of accuracy, the situation in existing society. Since Hume's day, a common complaint against contract theory has been that the myth of the original position exaggerates the freedom that human agents have to accept, or even choose, the social conditions in which they live. (For a brief account of the problems of applying a contractarian account of punishment to real societies, see Murphy 1973.) As will be pointed out in the next chapter, there are also questions to be raised about the notion of 'rationality' which looms so large in the Kantian moral ideal. For the moment, however, my purposes have been served if the cases discussed in this section have suggested extreme caution about simple utilitarianism.

In conclusion, I want to consider one final aspect of Hardin's account. He does not explicitly address the distinction between utilitarianism and contractarian alternatives, but he does try to suggest that the morality of an act is a question of *context*. He writes:

> Using the commons as a cesspool does not harm the general public under frontier conditions, because there is no public; the same behaviour in a metropolis is unbearable. A hundred and fifty years ago a plainsman could kill an American bison, cut out only the tongue for his dinner, and discard the rest of the animal. He was not in any important sense being wasteful. Today, with only a few thousand bison left, we would be appalled at such behaviour. (Hardin 1972: 256)

From the context principle suitable to such examples, Hardin goes on to suggest that the freedom to reproduce is, given the global situation in which we now live, intolerable. There is nothing right or wrong, in itself, with having children. By the context principle, the rightness or wrongness of certain acts is a function of the state of the whole system within which those acts are performed. Likewise, fishing the seas, spraying the crops with biocidal agents, playing a radio at full volume, and dozens of other acts are not morally assessable outside a context.

There is much that could be said about Hardin's contentions. For example, I would not agree that the example of taking the bison's tongue does not involve waste, however plentiful the bison were supposed to be. I have also modified the context principle to make it more plausible than the one he gives (which seems to apply to any act whatsoever). And I will ignore problems about act-identity, for example, that playing my radio at full volume in a crowded place might be said to be a different act from playing it at full volume in the seclusion of my home. What is important about the context principle, for my subsequent treatment, is its ecological nature, the fact that it indicates the need to think of decisions, attitudes, actions and feelings as located within a context. As will be argued in the next chapter, it is precisely the attempt to give a context-free account of morality that proves the ultimate undoing of both teleological and Kantian perspectives as they are normally deployed. What ecology shows is not simply that the context makes a difference to the kind of action we engage in. It shows, rather, that what kinds of things we are, what sort

of thing an individual person is, and what sort of options for fulfilment and self-realisation are open, are themselves very much
context-dependent. Only once this fact is recognised does it
become possible to construct a position that does some justice to
the ethical concerns urged by Taylor, Rolston and the deep
ecologists.

11.2 Prospects for environmental ethics

The argument in the last two chapters may have made the
prospects for any genuine environmental ethic appear poor. There
was hope for an extension of awareness and a reinterpretation of
the moral by virtue of the dignifying impact of ecology. However,
even though it is undoubtedly the case that studying nature has
made many professional biologists deeply concerned about our
treatment of the environment, our hopes for revolution in ethical
thinking were quickly dashed. Taylor's biocentric perspective
involved a distinction between value, worth and merit. But as
soon as the implications of Lewis's account of value were
considered, it turned out that the features by virtue of which
experience was intrinsically valuable, were precisely ones that
were not specific to the object of inherent value bound up in the
experience. In that form the theory did less than justice to our
conception of the significance of the individual, whether person,
plant or inanimate thing.

As an alternative to this experience-based version of value, we
explored Rolston's and Leopold's ideas about systemic value,
using the term 'value' this time in a way roughly synonymous
with Taylor's use of the term 'worth'. We had to face what is a
recurrent problem in the account of valuation — the difference
between our valuation of individuals as particular items of value
and of individuals as embodiments of clusters of more general,
community value. This time, the account promised to take us
somewhat further. Given that at least some of the things we value
are community values, then our ability to identity with wider
communities means that there is a perspective on which we can
view our own projects, behaviour and dispositions as instrumental
in promoting larger goals, goals of the social and natural
communities in which we are nested and which constitute the
home in which human activities are what they are.

However, no sooner had this account been mooted than an

obvious difficulty presented itself. Even if we can make sense, in the way Taylor suggests, of the good of individual creatures, sustaining their existence in the face of a co-operative, but often challenging, environment, ecology furnishes us with no 'objective' account of the goods, the ends or the directions of biological communities or ecosystems. As species populations come and go, as whole species die out and others emerge, there is no natural pattern, no direction, no end to be served. Rolston, admittedly, has argued in various places (for example, Rolston 1986: Essay 10) that evolution shows a certain direction — namely a tendency to produce a greater and greater diversity of species. But this is simply to associate a direction with what has in fact happened. Over aeons of time, more and more species have appeared. This observation by itself has no moral force, for just as the tendency of scorpions to maintain their lives against environmental challenges shows nothing about the moral worth of scorpions, so the tendency for species to diversify shows nothing about the moral worth of such diversity.

To balance the pessimism implicit in this review of our progress so far, it is worth recalling that the ideas of self-realisation and identification are not yet discredited. On the contrary, it is clear that many people do, in a certain sense, identify with nature. They feel upset, concerned, even affronted, by damage to natural systems, by destruction of rainforest, by pollution of the sea. Such identification is perfectly intelligible even if we have not yet uncovered its source. Likewise, it is perfectly intelligible to maintain that self-realisation requires not just social and institutional settings for humans, but a life involving certain relations with other natural things and perhaps an appropriate attitude to nature. The possibility of such accounts of identification and self-realisation involves a moral dimension, and thus provides evidence of the possibility of an environmental ethic.

In the event, I will be arguing for the claim that ethics is concerned with a mixed bag of principles, ideals, modes of deliberation, institutional structures, feelings and dispositions which probably defies systematic analysis. Thus, there will be no simple answer to be given to the questions: how ought one to live? In what does goodness consist? What makes actions right? A very strong tendency needs to be resisted: this is the tendency to think that there is one thing which makes actions right, or one source of moral value. My own view is that no such agreeably simple account of ethical life will be possible. We face continual pressures on our moral choices from diverse sources, and I will be content to show that there is at least an ecological dimension to ethics. It should also be clear, from our previous investigations into types of value, that the term 'value' itself is not very helpful. Again, there is a tendency to think that one characteristic — worth, value or whatever – will underlie all claims to moral standing and to moral considerability. This tendency is harmless if we accept the idea of value as a vague schema, an idea which can be implemented in different ways in different things. But, as with other matters of ethics, we have to be wary of thinking that value consists in the possession of just one set of characteristics.

Just as this manuscript went to press, I received a copy of Stone's new book (Stone 1987). It is gratifying to find so much by way of shared ground. He too finds modern moral philosophy besotted with simplistic theories of the moral which attempt to impose, in his vocabulary, a *monistic* account of rightness. He favours, however, a pluralistic conception of morality that is sensitive to the fact that evaluation of acts involves different frameworks from those used in evaluating agents, situations or institutions. My notion of ethical *polymorphism* (see Chapter 12.3) is not unrelated to his ideas on pluralism, although my concern in identifying the different polymorphs of the ethical is to show that the considerations that lead to respect for nature will be at odds sometimes with considerations regarding persons, our families and the kind of society we want to have. Where we are both agreed, and what the critique that follows is meant to show, is that no one systematisation of the ethical — whether utilitarian or contractarian — is likely to capture the richness, the difficulty or the subtlety of real moral appraisal.

11.3 Utilitarian and contractarian accounts

As seen in Chapter 10.1, it is necessary to distinguish between duties, respect and obligations owed directly *to* some item and those that are owed *regarding* it. For those committed to a conception of the moral that is essentially human-directed, our respect regarding nature is owed directly to other human agents. Although there are problems about specifying the class of agents in question — for example, if they are as yet unborn — it is a commonplace for conservationists to argue that they wish to protect some resources for the sake of their children and their children's children, and for preservationists to wish to leave much undisturbed that the conservationist would destroy. On this somewhat anaemic conception of environmental ethics, the direct objects of duty and respect are therefore other people rather than the land, the biotic community or individual natural things (Rolston 1986: Essay 1).

The notion of extending moral considerability to nature and finding values in nature sits unhappily with one very prevalent kind of moral thinking: this is the perspective of the utilitarian, the kind of thinker who, as we have seen, allocates moral value in terms of the overall benefit, or utility, of actions. Although the position is fraught with difficulties, it possesses a certain immediate appeal. It captures one important aspect of much of our moral thinking. Utilitarianism does this by being a *consequentialist* doctrine — one that focuses on the consequences or outcomes of actions as determining their moral value. It is also, more importantly, a *teleological* doctrine, which means that it sets out to define the notion of rightness in terms of the separate notion of what is good. Suppose we accept that satisfaction of desires, or of a certain kind of desire, is good. Then a teleologist would maintain that an action is right when it leads to desire satisfaction and that in general the right is what maximises the satisfaction of desire. By contrast, anti-teleological, or *deontological*, accounts of the moral deny that rightness can be thus defined. They will instead maintain either that goodness and rightness are not independent in the way supposed by teleologists or at least deny that rightness is to be determined in terms of maximising some good (Frankena 1963). On this conception of the moral it makes sense, for example, to maintain that it is right to defend individual freedoms or liberty even when such a course fails to maximise the total amount of desire satisfaction in a community.

Suppose, for the moment, that the consequences of our actions do, in some cases at least, determine their moral value, and let us additionally accept the teleological doctrine that what is right is what maximises the good, the latter notion being defined in terms that are independent of the notion of what is right. Now we can start to understand why the notion that natural things in general have value in their own right is not easily grafted on to such a moral position. It may appear easy to find a metric for weighting the consequences of action when the natural things in question are other moral agents. Like me, other agents have projects, interests, desires, pleasures and pains. Pleasure, and the absence of pain, are, as Mill observed, both desirable and desired, as is the ability to continue with projects once started and to satisfy other interests. It is tempting, then, to take such things as the presence of pleasure, or happiness and the absence of pain as defining what is good and thus as a means of measuring the moral value of an action's consequences. In classical utilitarianism, actions are right when they maximise pleasure overall, or at least minimise pain. The value of an action, as Mill again noticed, is thus independent to some extent of the worth of the agent. Evil intentions can — given the ups and downs of life — lead to better consequences and more good on some occasions than the best will in the world. An act performed with the best of intentions can, likewise, go disastrously wrong and lead to far greater pain than some alternative.

By contrast with teleological thinking, accounts of morality that concentrate on the nature and personality of the moral agent are more sensitive than utilitarianism to the familiar fact that even when things turn out in unforeseen and unintended ways, we are often concerned with the agent's intentions and motivation when we give our verdict on the morality of what has been done. We also care about the agent's degree of concern for the rights and liberties of others. The focus of our concern is thus less the goodness of the state of affairs resulting from an action or set of actions, but instead the moral integrity and decency of the agent. On such an approach, we can also start to make sense of our attitude to distasteful means, and 'dirty hands' problems. A conception of morality which focuses particularly on the overall maximisation of benefit is apparently compatible with permitting injustice and pain to innocents where this is an essential condition of maximising overall utility. Introductory works on utilitarian ethics are replete with examples which show the implausibility of crude act-utilitarianism in the face of such problems. Utilitarians have the option of declaring that it is the principles upon which we act, and not the specific outcomes, that have to be evaluated by considerations of utility. This revised position, known as *rule-utilitarianism*, is itself open to further objections — a common one being that it collapses back into act-utilitarianism.

Given these problems, and given that there are alternative moral theories available in the form of Kantian or contractarian positions, we might wonder why utilitarianism retains its appeal. There are two reasons, I suspect, for this. One is that, as already noted, it does justice to the fact that some of the time, consequences (or expected consequences) do matter in moral thinking. This consideration alone is not very powerful, for it is a simplistic deontologist who regards situations and the outcomes of actions as morally irrelevant. There is no need for deontologists to take such a line. Indeed, Rawls dismisses such versions of deontology as 'crazy', maintaining that any moral position must take account of the consequences of actions in its assessment of their rightness (Rawls 1972: 30). A second, more powerful, reason, is that utilitarianism lends itself to theoretical development in a way that other moral accounts do not. Indeed, if we are to use the term 'theory' in connection with morality, then utilitarianism comes closest to deserving the title or moral theory. Not only does it provide a principled means of arriving at decisions in the face of moral problems, but since it deals in what may seem to be in principle measurable utilities (like pleasure and pain) we have the prospect of using mathematical and graphical models to help us in our moral deliberations. The utilitarian can try to provide a sort of economy of morals in conjunction with the rest of the theory.

It is not surprising then that some recent thinking about our relations with other animals and with the environment in general has been undertaken from a utilitarian perspective (Singer 1981; Attfield 1983; and contrast with Clark 1984 and Regan 1981). The extension of moral consideration to all the higher animals, is — on classical utilitarian grounds — perfectly sensible. Few people these days would try to deny that some animals other than humans are sentient, and so there is no good ground for discounting their pleasures and pains in any moral calculations we make. Of course, there will be problems about determining the degree to which the pain of a non-human animal is like the pain of a human. Perhaps the fact that we are rational, that we can reflect on the significance of our pains, that we have the foresight to see things getting worse, and allied considerations make pains much worse for us than they are for other animals. But, however we resolve that particular issue, there is still no reason to apply such a high discount to the suffering of other animals that it fails to figure in our calculations at all.

But if sentience provides us with grounds for allocating interests to other creatures, it goes nowhere towards providing grounds for allocating moral consideration to those creatures and living things that are not *sentient* at all, nor to those individual things or systems that strike us as being valuable despite the fact that they are not *living* at all. In the face of these sorts of cases, a

Kantian or a contractarian account of morality does not fare much better. To follow the Kantian injunction to act in such a way that the principle on which we act would be adopted by any rational agent in similar circumstances leaves us in an embarrassment when we try to spell out to some sceptical — though apparently rational — critic our reason for sparing the tree, respecting the integrity of the valley or trying to maintain some threatened corner of rainforest. Not every rational agent in the same circumstances need feel obliged to care for such things, even if it be admitted that such moral concerns are rational in the first place.

The contractarian account of morality sketched on page 26 in Chapter 2.1 has received some enthusiastic support recently. Although the original impact of contract theory was in explaining the political coherence of social groups, it is not hard to extract a kind of moral fable from the typical contractarian story. To recap, we are to think of our society as having come about due to an imagined contract whereby the members of society variously surrender certain freedoms in order to enjoy the protections and benefits of communal living. Of course, the benefits of life in society are not equally shared out: as is obvious, all societies contain inequalities, and in none is there much prospect of resolving all conflicts of interest and value. On a liberal contractualist view, such a result is unsurprising; for the surrender of freedoms necessary for social living is the minimum compatible with reaping the rewards. Now think of the social contract in moral, rather than political, terms. We live according to quite widely shared conventions of behaviour, conventions which secure certain rights and freedoms only at the expense of limiting other rights and freedoms. To live with others in a society, we might argue, is tacitly to accept a kind of contract concerning the bounds of permissible behaviour.

This mode of thinking associates naturally with a certain view of the human person as an *autonomous* agent. Such agents, as persons, are all entitled to respect, and are the sorts of things which have rights. The basic rights of such agents are all on a par, and thus any community of such beings must respect their fundamental rights to live and flourish in their own autonomous ways. Rawl's version of contract thinking asks us to imagine an 'original position' described as follows:

. . . no-one knows his place in society, his class position or social status, nor does anyone know his fortune in the distribution of natural assets and abilities, his intelligence, strength and the like. I shall even assume that the parties do not know their conception of the good or their special psychological propensities. (Rawls 1972: 12)

The original position is meant as a purely hypothetical device to let us think about choosing principles of justice suitable for organising autonomous beings into a society. The principles are chosen behind a veil of ignorance, so that no agents are biased by knowledge of their own, or others', specific talents, abilities or disadvantages. A further point is that any principles chosen by our hypothetical and ignorant agents should be taken as chosen by *rational* persons. Whatever the autonomy of persons is meant to amount to in the end, it is clear that, as in Kant's earlier defence of the view, rationality is an important component (Brandt 1979; Diggs 1981).

Relative to the original position we can now consider the question of how fair a given society is. We can ask whether its principles, its ideals of justice, its moral and legal codes, are of the sort that could have been rationally chosen by agents in the original position. Think of contemporary South Africa, and consider a hypothetical original position in which the members of its present population would have chosen the rules and principles of a just society. It is not easy to square the present system of apartheid there with any conception of rational choice behind a veil of ignorance. On the contrary, apartheid is a body of principles and doctrines which could only have been advocated by a group that had reason to believe that it would occupy a position of advantage under a social contract that institutionalised racism. Put another way, the apartheid system is a clear example of social organisation that fails to treat people equally in respect of their fundamental human rights; that is, it fails to recognise the autonomy of the person.

The central idea of autonomy is not itself easy to define. Different thinkers have used the notion in slightly different ways and I have elsewhere queried the idea that the notion is exclusively applicable to rational beings (Brennan 1984). The basic idea is that human persons are self-controlling, self-governing beings, whose growth and development is subject to higher-order constraints that are themselves internal to the person. For example, I not only have desires, but have desires about my desires, as well as desires about the sorts of ideals, virtues and dispositions I would like to embody. I not only value, and take up attitudes to the things around me, but evaluate, and take up attitudes to my own values and attitudes. Since autonomous beings are self-choosers as far as their own goals and ends are concerned, morality cannot lay down precisely what ends persons should choose. The point of the choice behind the veil of ignorance is that it forces us to think of morality as licensing the ability and freedom of rational beings to choose their own ends, subject simply to the constraints imposed by the fact that other agents will also be engaged in the same project. (For subtle and interesting discussions of the contractarian position, see Richards 1981 and Williams 1985.)

The self-conscious, self-critical features of the person emphasised by contractarian accounts give some substance to the notion that human pains, desires, needs and projects are of a much richer sort than the comparable features of other animals. If there is something especially precious about human life, a value that is lacking in other forms of life, we can perhaps start to articulate what this is, along the lines just sketched. Certainly, for any kind of contract theory of the moral, the points about human autonomy suggest that there is no prospect of arguing that the original position or state of nature should encompass a community of humans, other animals, living but non-sentient things, natural systems and rocky crests as participants in the social contract. If we are to argue for an extension of ethics that takes us beyond the social contract as conceived by Rawls, then we cannot do this simply by adding to the list of original participants. The prospects for an environmental ethic based on contract theory thus look bleak. The utilitarian can, at best, count in some non-human animals as having pains and pleasures that are worth considering in the operation of our calculus of goods. The deontological alternative may likewise allow a special class of mammals or higher animals to be autonomous to a degree (though the extent of that must be debatable) and to that extent worthy of moral consideration. Even if neither conception motivates the claim that some animals other than humans have rights, there is the possibility of arguing that there are living, sentient things other than human persons that merit respect. But all this is to go no way towards finding principles of a morally compelling kind that will restrict our actions towards other natural things, except where the interests of other autonomous beings are concerned.

As already suggested, there is a great deal of evidence, much of it supplied by scientific ecology, that it is in the interests of autonomous agents (however the class is delimited) to treat nature with a higher degree of respect and concern than is currently exercised. But what is more challenging is to try to outline a conception of the moral that will do justice to the idea that there should be more to an environmental ethic than injunctions flowing from this sort of self-interest.

12

Beyond the Social Contract

12.1 Ethics beyond utility

One of the central pieces of ecological insight — whether of scientific or metaphysical ecology — is that each thing is *what* it is in part by being *where* it is. The formulation of the insight is vague enough to cover multiple instances. Living things are complicated arrangements of only a small number of elements from their environment, and continued existence of each depends on being located in the right place on the great cycles of matter — the nitrogen cycle, the citric acid cycle, the carbon cycle, and so on — which enable elements to be used, and re-used by several living individuals. In this way individual things 'suck' (in Schrödinger's phrase) low entropy matter from their surroundings and sustain themselves by it before returning it, ultimately, as high entropic waste.

But living things likewise bear the marks of evolution — having morphological features, and behavioural traits critical in the past and still essential for the survival of the genetic material they carry. Each of us thus carries the marks of the past evolution of our species. At a more mundane level, the use of resources by an organism may depend on where it finds itself: its body chemistry, its weight and its predatory role thus reflecting its position in some particular ecological locale (bear in mind the distinction between fundamental and realised niche). These are all aspects of the truth which we saw articulated in the third chapter, that organism and environment are complementary, each inseparable from the other. We need not go to the extreme lengths of those who, like Lorenz, describe the hoof of the horse as 'mirroring' the steppe, in order to get a feel for the point.

At the level of society, human groups and human institutions, the same truth receives clear illustration. The redundant worker, embittered by early loss of work, can in no way be understood separately from the social circumstances which produced the problem situation. Apart from the institutional practice of thinking about and writing on philosophy, my activities in composing this book cannot be explained or understood, nor therefore those aspects of my life that involve commitment to philosophy. Notice that recognition of the ecological point in no way subverts claims about the individual autonomy, worth, freedom or dignity of persons as agents. I am not saying that redundant workers have no choices over future life strategies, nor that commitment to philosophy leaves me no freedom to negotiate other aspects of my life. But, as I will argue, freedom and self-regulation always involve constraints. Indeed, a plausible diagnosis of our current environmental problems is that we have not taken the constraints on human choices nearly as seriously as we should.

I now want to show that modern ethical theory in the form I have described it suffers from ignoring ecological facts of life. In brief, my complaint is that notions like *good*, *rationality*, *interest*, *desire*, *obligation*, *freedom* and the other key notions of ethical theory have been applied in ecologically careless ways. Theorists have taken little care to make sure that these notions are realistic for creatures that live in real social and biological surroundings, and are what they are by virtue of their location in such surroundings. The duality of value (described in Chapter 10.3) and the apparent puzzle it poses become no longer puzzling once a suitably ecological approach to ethics and value is adopted.

Consider, for a moment, teleology as an ethical strategy in general. The idea, recall, is to give an account of right conduct and virtuous action in terms of some further, separate notion of the good. Once we know what state of affairs would be good, we can take steps to act so as to secure it — action which will, then, be right. In its utilitarian forms, as we have seen, the notion of good becomes further specified with respect to happiness, needs, desires or welfare. But what is totally artificial about such an approach is the notion that somehow what is good in terms of welfare, utility, or whatever, can be defined independently of all social, biological or economic situations whatsoever. It is highly doubtful, in my view, if we can give much content to the notion even of basic human needs, independent of any context at all. Certainly, if we do think about this in the abstract, then all we can say is that humans need to live between certain minimum and maximum temperatures, to have sources of fresh water, some sort of food, and some sort of social context in which to flourish. In ecological terms, we can indicate that the *fundamental niche* of the human species is extremely broad (see Chapter 4.1) — and this is true in literal, biological terms, as well as being true, figuratively, of the social, cultural and political lives of humans. So the realised niche of one population can vary widely from that of other populations.

Consider a life of the sort dismissed by Kant in a notorious passage (Kant 1785), the life of an imagined South-Sea islander with simple pleasures, a gentle climate, rich fisheries, ample supplies of naturally available fruit and root crops, and little by way of natural hazards. However idealised the picture, we can at least make sense of a form of life that is gentle, pleasant and comfortable in this way, supported by relaxed social arrangements but with enough by way of ritual, other communal activity, sport and seafaring to make it a life worth living, a life that can be lived, relative to the context, in better and in worse ways. Now once we enter this mythical context, we can start to get hold of the idea of various kinds of social and perhaps even economic goods. Suppose our islanders have no money, but nonetheless engage in activities relevant to displays of wealth and status. The accumulation of objects of various kinds, or the ability to offer feasts to guests, will perhaps be goods that at least some members of the society strive after. On the other hand, if the society is governed by a ruling clan, or a council of elders chosen simply by age, then political aspirations will just not be possible for many of the society's members. The point is that what counts as desirable, or as contributing to welfare or happiness will be constrained by the actual conditions of life in the society, and these in turn are constrained by the environment of the islanders (in the imagined case, a relatively bounteous one).

Clearly, the basic needs of the imagined islanders are quite different in detail from the basic needs of Eskimos. Take the need for warmth and shelter, for example. Our islanders may require little more protection from the elements than is necessitated by the customs, taboos and rituals of their community, while survival inside the Arctic circle demands heavy protective garments and insulated, weatherproof shelters. Of course, we can insist that, in some vague sense, both lots of people have warmth and shelter as basic needs. But such insistence is just as empty of real content as

Mill's claim that pleasure and the absence of pain are the fundamental goods. For, in the sense in which this is true 'pleasure' is simply a cipher term, for which we would need to make substantive substitutions in real social contexts. The ecological and the social points here are parallel. Just as the total natural environment determines that certain options are not available (nakedness, or fruit-gathering for Eskimos) so a social environment rules out certain kinds of aspiration, while directing attention to others.

I am not arguing for any kind of extreme conservatism here. The fact that the actual conditions of life within a society put constraints on what aspirations are possible for those who live their lives within it, does not mean that there are no prospects for changing society. On the contrary, every society has its own dynamic, and within it there will be currents of change as well as currents of tradition. How these forces combine to bring about the creation of new social structures over time depends, in my view, in part upon the contributions of individual persons, of groups of persons and on social and political forces that have their origin outside the society. Nor do I deny the possibility of criticising a society and its form of life, either from the outside, or from within, by bringing forward considerations about rationality, dignity, respect and moral decency which are drawn from a fairly remote standpoint where human life has been considered in universal and highly general terms. What I do argue against is *ethical colonialism* whereby those situated in certain kinds of society, surrounded by certain kinds of goods and activities, declare that they have discovered universal ethical truths which do no more than reflect, in a suitably generalised way, their own local aspirations and ideals.

It is hardly surprising that a degenerate form of utilitarianism has become the unofficial ethical stance in many of the highly developed, rich countries of the world. Anyone born into such a society is likely to acquire conceptions (that are by no means universal) of what is desirable in life. Yet within such societies, expectations are disappointed in those who lack the material goods that are widely available (though only at a price); moreover, market forces, if operating free of ecological and moral constraints, can ensure that demand be created where there was none before, and put ever more appealing lures before consumers. To maximise interest and happiness can easily be translated in terms of maximising monetary gain in a way that enables people to afford the multitudinous goods such societies offer. But now the position being advocated is morally irrelevant. For maximising utility is merely the maximising of self-indulgence. Without restraint on the pursuit of individual satisfaction there is the risk of exceeding the capacity of the society, and the environment on which it depends, which is then disastrous for the society as a whole. The disastrous results of exceeding biological carrying capacity are well illustrated in Hardin's fable and in the real life destruction of common agricultural, forest or fishery resources under the impact of unbridled exploitation.

One serious challenge facing the utilitarian, then, is how to refine the theory to avoid the tragedy of the commons and the advocacy of lives that are shallow and ultimately unsatisfying. Utilitarianism suits the liberal conception that each of us is able to define his or her own good in terms of expectations, utilities, goals and projects that have value for us. But, as Mill clearly saw, any untrammelled version of the theory thus opens the door to the claim that a life of trivial pleasures is superior to a life of frustration where the agent has been intent on serious, but difficult, projects (Mill 1910: Ch. 2). Moreover, the more we pursue individual good, conceived as individual interest, the greater the prospect of bringing about harm for the community in just the way Hardin's imagined commoners end up destroying their mutual resource. Mill — in a controversial but consistent passage — identifies the general happiness sought by utilitarians with individual happiness:

> No reason can be given why the general happiness is desirable, except that each person, so far as he believes it to be attainable, desires his own happiness . . . each person's happiness is a good to that person, and the general happiness, therefore, is a good to the aggregate of all persons. (Mill 1910: 32–3)

The 'therefore' in the above passage has been much criticised, but it is hard to see what other way out there was for Mill. His problem is that he seemed to think that two potentially incompatible goods could both be maximised — the good of the individual and the good of the group. Without further modification, or the addition of information about linkage between the two goods in question, it is hard to believe his conclusion is at all plausible. Any moral doctrine that urges the pursuit of individual welfare needs, if it is to be taken seriously, to be accompanied by an account of general constraints that will prevent the egotistic pursuit of benefits from crippling the community as a whole.

Utilitarianism thus does not fare well, and — if we take it as representative of a powerful strand of teleological thinking — we are liable to be suspicious of the ultimate prospects for any decent teleological theory that is not ecologically motivated. One route that is possibly worth exploring would make use of a distinction, originally due to Rousseau, between the good of all and the communal good. Although we might try — as Mill appears to have done — to think of the good of all as the sum of all the individual goods in a community, the communal good might be quite different. The communal good would be so defined that destruction or subversion of the institutions that support individual goods would be contrary to it. What our teleological perspective would then deliver, in theory, is an account of how communal goods can be achieved compatible with maximising individual freedoms, and those aspects of welfare that are compatible with sustaining the society in question. On this perspective, the ethic of a society would have to be more ecologically based than modern ethical theory seems to have recognised.

12.2 Persons, primary goods and duality

If classical utilitarianism is a clear example of a moral theory that fails to be ecological in the required way, so are most developments of the contractarian perspective. Indeed, this is hardly surprising given the formal point that utilitarianism is but one very special kind of contractarian theory (Williams 1985: 81 and references to Harsanyi's work). The theories, however, fall short of the ecological ideal in slightly different ways. Whereas, we might say, the utilitarian thinks of welfare in society in general, in abstraction from any particular social arrangements, the contractarian focuses attention primarily on moral agents in complete separation from the society in which they are supposed to live. Such separation is a condition of the thought experiment proposed by Rawls and which defines his 'original position'. Notice that it is not the attempt to say something about the commonalties of human nature and human society that is objectionable. In each case, the trouble is that the theories try to give an account of persons who live in society in a way that ignores the force of the claim that what I am is a function of where I am.

A fundamental notion of contract theory is that of *rationality*. The agents who choose principles of justice for a society behind the veil of ignorance do so as rational beings. But the very idea of rationality depends, for substantive content, on a social context. What seem to those who live in the Anglo-American intellectual tradition to be hallmarks of rationality — due deference to authorities of one sort, scepticism about certain ranges of belief, acceptance of a wide range of 'common sense' beliefs — are very much an artefact of their historical position, or even their position within an individual life.

Another key notion in contract theory, as we have seen, is the idea of autonomy — that self-regulating freedom which rational agents are supposed to possess. But no living thing, if it is to survive, is free in any absolute sense. Without lapsing into determinism (see Chapter 1.3), we can recognise that physical and social environments put constraints on our actions, and that choice occurs within such constraints. This means that Rawls' account of *primary goods* has to be taken with some scepticism — or be treated as lacking substantive content.

The primary goods he mentions are those which agents in the original position will want if they are going to want anything at all. The goods are rights and liberties, opportunities and powers, income and wealth, and a sense of self-respect. As soon as we consider which rights or liberties, which responsibilities or powers agents will want, we can only make progress if we relate our thinking to some social situation. But to do so is to think of a case where there is no longer the ignorance assumed by the myth of the original position. As Bernard Williams has pointed out, it is not obvious that wealth or income would even be wanted in many possible social situations (Williams 1985: 80). We cannot realistically compare justice in actual societies with choices about justice and other goods made in a situation that so falsifies the real one. The contractarian — just as much as the typical teleologist — gives us one side of the ecological story and then tries to build ethical principles while ignoring the rest.

It is no wonder then that regular ethical theory has proved unhelpful in dealing with the problems of natural value and living in a natural community that have so exercised us. If we are to develop an ethic that is suitable for the real circumstances of our political, economic and physical situation, then this will have to satisfy at least the ecological adequacy condition. This condition stipulates that the ethic take account of the dynamics of persons — who are, after all, the only moral agents of whom we so far have knowledge. By the 'dynamics' of persons, I mean simply that persons grow, mature and live their lives in particular natural contexts, at particular places in history and within specific cultures and societies.

We can give some more content to the adequacy condition by going back to reconsider a puzzle which bothered us in Chapter 10.3. One aspect of the puzzle was that to some extent we seemed to value persons for what they are in themselves, and to value other things by association with the valued person. At the same time, we also valued persons for what they did, for the values they represented, carried on and passed on to others. In the second case, persons could be replaced by substitutes and yet the same valued things would go on. In the former kind of case, by contrast, part of the value of an activity I do with someone involves that person's irreplaceable participation.

The ecological solution to this particular puzzle is to allow both sides of the valuation to carry weight. Persons are what they are by doing what they do, having the friends they have, forming the relationships they do, carrying out various projects, undertaking various commitments. Looked at from the side of the person involved, a life is a dynamic process of thinkings and doings which involve what Dewey called the 'continuous reconstruction' of experience. More recently, Wollheim has drawn attention to this aspect of self-development (Wollheim 1984), showing how echoes of previous experience resonate with present experience to make it what it is. My current experiences, then, would not be what they are were I not living at this time, in this society, of this species and carrying within me a distinctive history of previous experiences. To be me, then, involves the continuous reconstruction of this, my life, through me-ish experiences.

The other side of the same coin, however, is that my experiences and doings essentially involve the social and physical world that I inhabit. My friends, relationships and environing objects have played their part in constituting my 'identity' (for want of a better word). I do not mean that they literally constitute me, for that would be too strong a claim and lead me into a theory of the person as social construct. Rather, I mean that these things are what give my experience, modes of response, sense of humour, and so on, the very distinctive character that mark them as mine.

180

None of my experiences are unique to me: others could have had some of the same ones, and no doubt many do. But the peculiar range of contacts, associations, and experiences that chart my path through the world around me will be something that is very likely unique.

The duality of agent and deed that seemed so puzzling now reappears as an important truth about our extended conception of personality. We are carriers of history, and that means carriers of marks from our own past and from the world around us. No wonder it seemed strange in a way to think of persons as having value in themselves — for the notion of 'self' implicit here seemed forever to elude proper description. Yet at the same time, persons are more than mere carriers of valuable views, doers of worthwhile deeds and furnishers of worthwhile experiences. To take persons as merely the latter kind of agent seems to diminish their agency and leave us no other form of value for them but the instrumental. What I want, however, in sharing an experience with a friend is precisely the combination of that friend's unique participation (which will colour the experience in a specific way) and those responses, valuations, jokes and the rest that are distinctive of — but not unique to — my friend.

The duality of which we have now made some sense is simply mirroring at a higher level a duality already present at more fundamental levels. Consider the famous nature–nurture debate. An organism arrives on the earth equipped with a genetic program, a program that will determine, among other things, the size, life-expectancy and fitness of the very body on which the program is to run. But the program is not the sole determinant of such things. Genetic potential interacts with the actual circumstances of the organism to produce the outcome that is an actual life. Thus a trout in an overstocked Hebridean loch will grow to only a fraction of the size of a trout in a stream in which it has less competition for less scarce resources. Perhaps the trout in the stream is flourishing to a greater extent than the one in the loch; this, however, is debatable. But within each environment, and relative to each fish's genetic potential, there are possibilities of growth and flourishing, as well as possibilities for harm. Each trout, in whatever environment, is equally a member of the species *Salmo trutta*. That is its nature, or kind. But what it is, what it becomes, how it preserves itself in response to the challenges of its environment is not a story exhausted by the bare information given by taxonomy. The same is true of humans, with due allowance for the greater richness of the story about our own lives.

12.3 Living in nature

Let us review what has been suggested in this chapter so far. Neither teleological nor contractarian theories, as standardly developed, have taken on board the ecological point that what we are depends partly on our physical, social and historical location. The normal versions of these theories thus have at most a kind of formal plausibility (as when we are exhorted to pursue happiness for everyone, when the really interesting ethical work to be done involves determining just what styles of happiness are possible, fulfilling and sustainable in real social and physical situations). It is not clear that the answer is to recast the theories in new forms that take account of the interdependence of agent and environment. In the case of contractarian theories, for instance, this recasting seems doomed to failure. For such theories require choices from a position in which all social and physical constraints are discarded.

It might be possible to introduce choice behind a veil of partial ignorance — where we are given information about the unsustainability of various forms of social life, in the light of certain information about the physical, environmental constraints. Thus a group of rational Eskimos, a group of rational South-Sea islanders, and a group of rational west Africans might come up with quite different conceptions about what would constitute a just society in their specific circumstances. But if we are allowed to build in this much information, why stop simply with the environmental facts? Once we depart from the conditions of initial ignorance, it is hard to know where to stop.

With teleological theories, there is also a difficulty. Any account of goods, welfare, or benefits that are to accrue to people as a result of their ethical decisions has to answer to the objection that not all that seems good is good, that what once seemed good to me no longer seems so, and that many things we want, like and even claim to 'need' may be incompatible with the sustainability of the form of life required to produce them. The average child in the United States consumes 40 or 50 times the resources of a child in an economically poor country. The prospects of bringing all children to the astonishing level of consumer satisfaction of those in the rich countries is not compatible with maintaining life on earth. Given these stark facts, any unrestricted appeal to interest, desire satisfaction and maximisation of personal goods looks pretty sick as the basis for an ethical theory. In order to lick utilitarianism into good shape, then, we may have to start thinking of the ethical community over which it operates as the global village, recognising that wealth and indulgence in one part of that community can only be bought at the expense of degradation and poverty over the rest. It is interesting to note that some contemporary utilitarians have taken this very route — thus committing ethical theory to a scope that it has great difficulty justifying (Singer 1972). For a start, I have enough trouble trying to sort out the implications of my choices for those near, and sometimes dear, to me, let alone working out their remote consequences.

What has gone wrong with modern ethical theory when its two most prevalent forms can be attacked so readily? The ideally rational agent of the original position is too shadowy a beast to be recognisably human at all, while the eager pursuit of the utilitarian paradise leads all too quickly to self-indulgence and extravagance. Two things have gone wrong. First, neither account makes contact with the real position of human beings, since neither takes very seriously the idea that we have a nature by virtue of which our potential is capable of better or worse fulfilments. Second, as I will argue at the start of the next chapter, there is no prospect of deriving a single, unified ethical theory consisting of a set of principles that legislate for every possible situation. This is not so much because of the richness of possible situations but because of the fact, already emphasised in my treatment, that human life is itself a complex phenomenon which can be viewed in terms of infinitely many frameworks. There is no reason to believe that what appears the right things for me to do as a member of my family will coincide with what may be the right things to do as a member of my university, of my profession or of my species population.

Between them, these two points are fatal for the moral enterprise as it is commonly conceived. Since more will be said about the second consideration in the chapter that follows, I conclude this one by clarifying my stand on human nature. It is important not to confuse the position I am supporting with two alternatives. One is the view that there really is no such thing as human nature — that human beings are almost infinitely plastic, so to speak. On this account, each individual defines a set of projects, or perhaps a fundamental project, in terms of which that individual's identity is forged. 'There is naturally an infinity of possible projects as there is an infinity of possible human beings' (Sartre 1958: 564). The ecological position that I adopt accepts this view as only a partial truth. Existence does, as far as some of our significant properties are concerned, precede essence. But our history of adaptation, and the offerings of our environment, set limits to our freedoms and our projects, so that some are apposite to us in a given situation and others are not.

The other position from which mine must be distinguished is the classical 'humanist' position of many of the great thinkers and religions. On these accounts there is such a thing as human nature, there are prospects for life in keeping with that nature, and traditional theories provide what is put forward as *the* truth about that nature. But I deny the widely-shared doctrine that the whole truth about our nature is knowable, or can be stated within some framework or scheme. It should be clear from the opening chapters that I regard any attempt to capture the truth about human nature in any scheme or theory as almost certainly vain. The classical theories of human nature were correct, however, in recognising that human nature is not infinitely plastic, and that the question of what makes for a fulfilling human life is to be asked in the context of some account of our natures.

My 'ecological humanism', to give the position a name, urges that human nature is tremendously rich and complex, for it is partly what we are born with and partly what we make for ourselves as we live. For humans, like all other natural creatures, grow and develop by interacting with their various environments (social and natural). Unlike other natural creatures, we also have a highly developed ability for self-reflection and self-awareness. But what is this 'self' upon which we reflect? It should now be clear that it is not simply something with which we are born; rather, it is the continuously reconstructed self which to some extent reflects the various domains which the subject inhabits. As someone who lives in a capitalist economy, for example, I can reflect on my economic life, and the ways in which my desires, wants, interests and pastimes are reflected in my engagement with the world of the economist. If I were an inhabitant of a very different kind of society, there might be little, or no, possibility of engaging in similar reflections.

It has been seen that there is little by way of detailed content that can be given to the idea of human needs outside the actual situations in which humans find themselves. The idea of human wants is likewise relative to time, place and context. To say this is not to say that any wants engendered by a particular context are ones that ought to be pursued. When considering a society and its political and economic relations we can use ecological analogies to criticise the forms of life or role models that are widespread in it. But it would be wrong to infer from this that there is some unique, specially satisfying set of wants that it is desirable for human beings to cultivate given their nature. The ecological account of human nature allows for both richness and complexity as well as for diversity. But it involves the recognition that all human life is lived within some natural context and that it is in terms of that context that the identities of very different human lives are forged.

13

Living in the New Community

13.1 Ethical polymorphism and environmentalism

It has already been suggested that an ethic by which to live is not to be found by adopting one fundamental, substantive principle relative to which all our deliberations are to be resolved. Instead, we are prey to numerous different kinds of consideration originating from different directions, many of them with a good claim to be ethical ones. In any actual situation in which we have to make important decisions we will also have other factors to weigh — non-ethical ones which may be significant in affecting our choices. For the romantic, it might seem that love requires, at least on occasion, that the commands of morality be ignored. And there is certainly nothing unintelligible about the claim that, however serious, the dictates of morality are not necessarily overriding. What will be considered in this section, however, are not factors on decision-making that might be hostile to the moral ones. Rather, I want to take seriously the idea that morality is *polymorphic* and that within its complex structure there will be room for ecological or environmental considerations.

Among the writers so far considered, several do seem to recognise the polymorphism of ethics. Taylor, for example, avoids the temptation of trying to reduce all ethical considerations to those arising from respect for persons. Instead, he takes such respect as one principle, but one that has to be integrated with the much wider principle of respect for nature. Since persons regularly have an interest in doing things that inevitably disrupt or interfere with nature, there will be a need for priority principles to determine the legitimate scope of such interference. Taylor identifies five such principles, and I will briefly review them in order to give an indication of the kind of ethic he commends. His central notion of impartiality between the claims of living things, no matter how sentient, how intelligent, or how vocal they may be, might seem to pose a difficulty for us in determining how to deal with disease organisms, or with creatures that threaten our lives. However, an impartial principle of *self defence* allows that any species or member of a species, can take action to defend itself from any organism that threatens its basic health or well-being. For us to take such action, though, the threat must be real, and there must be no alternative means of protecting ourselves available.

But what if we want to destroy living things in order to make a beautiful garden, or build a theatre? For Taylor, such situations must be considered in terms of principles of *proportionality* and *minimum wrong*. The first of these principles gives more weight to basic interests than to non-basic ones, while the second insists on limiting damage done in pursuit of non-basic human interests in cases where this damage involves the basic interests of other organisms. Thus theatre and library building, gardening and industrial development can all be ethically carried out (in accordance with the principle of minimum wrong) but also a whole lot of others count as unethical. Interestingly, Taylor makes no distinction between activities like fishing for sport on a modest scale, and slaughtering herds of elephants with machine guns in order to poach ivory. Both are equally cases of exploitation ruled out by the proportionality principle: the interests satisfied by hunting and poaching are non-basic, while the interests of the creatures killed are basic. Taylor believes that to satisfy peripheral interests in this way at the expense of the basic interests of other creatures is intrinsically incompatible with respect for nature (Taylor 1986: 274–8). As with other writers in this tradition, and along with some animal liberation theorists, Taylor allows that purely subsistence hunting constitutes an important exception: it is not exploitation, but licensed by the proportionality principle since the basic interests of the subsistence hunter are at stake.

It is clear, I hope, that Taylor's principles raise more disputes than they settle. How are we to tell when some non-basic, or peripheral, interests are worthy of interference in the basic interests of some organisms? If we take the claims of all living things as being equal, then on an impartial view it is hard to see that we could ever justify sacrificing the basic (life or death) interests of any for the peripheral interests of others. However, even if we think that some non-basic interests do trump basic interests, we still face a problem of specifying and defending such choices. Taylor clearly believes that building libraries and theatres is important enough to justify the deaths of millions of organisms, while fishing and other sports are not. But who is to say that literature and culture are more important than fishing? An academic, of course, with a vested interest in 'high' culture would possibly say so; but this may simply ignore the significance of fishing to the lives of other people.

The same kind of problem arises for his fourth principle, that of *distributive justice*. According to it, if satisfaction of basic interests requires the use of the same, limited resources, then these resources should be apportioned fairly. This leads to an immediate conclusion in favour of vegetarianism: if it takes seven tonnes of cereal to produce one tonne of beef, then clearly meat-eating uses up resources shared by humans and other creatures in a way that is quite unnecessarily wasteful. The less land used by humans, the greater the prospects for the flourishing of other forms of life. But the confirmed meat eater may consider that it is perfectly proper to reduce resources available to other life forms for the sake of certain exquisite gustatory delights.

To criticise Taylor's principles in this way is not meant to suggest that they are without merit. In fact, if we made them context-relative, they would increase immensely in their plausibility. As I have argued, the content of human basic needs depends very much on environmental factors, and an extension of this point involves recognising the context-sensitivity of non-basic interests. So in trading off basic interests of animals and plants against non-basic interests of humans, we can give consideration to the circumstances of the society or group of humans and other living things in question. A culture with no commitment to sport, after all, can hardly claim to need a football stadium, while, in European, North American and Australian societies, such projects may legitimately claim a reasonably high priority. What we need to be wary of, in all these cases, is letting our pre-theoretic notions lead us astray. For our intuitions will themselves be formed to some extent by our own social, environmental and institutional situation. A properly ecological development of Taylor's principles may give them more chance of gaining acceptance.

What of harms already done to plants and non-human animals, or that we may be forced to do to them, even in the course of projects carried out in respect of Taylor's other principles? His

final principle, that of *restitutive justice*, requires that we try to
compensate other organisms for harm that we might have done
them. Strictly speaking, compensation is not possible at all in
some cases. For if I kill an organism, it is no compensation to it
that I have set aside a protected area where others of its kind may
live. Respect for living things, of the sort urged by Taylor, seems
to me to sit uneasily with the doctrine of compensation, although
we can perhaps make sense of this in terms of wider notions, like
Leopold's, of respect for the biotic community. As noted earlier,
Taylor's own position seems to force him to maintain that
communities and ecosystems are themselves inherently worthless
and thus outside the scope of the ethics.

The objections made here to Taylor's proposals show how easy
it is to be sceptical about any set of ethical ideals. Scepticism is
very much a feature of contemporary western society, and it may
be that moral scepticism is a bad thing. There is more than a little
truth, I fear, in the complaint that the demise of religion has been
accompanied by a loss of moral conviction. But the kind of objections just made to Taylor do not stem from any global scepticism
about the moral enterprise on my part. On the contrary, I share
with him a concern to articulate a point of view relative to which
many of the acts we engage in, many of the attitudes we strike,
and many of the dispositions we inculcate into our children are to
be seen as wrong, as ethically defective. But although Taylor may
be moved by a love for, and concern about, nature, he has failed
to come up with principles that are grounded in ecological reality.
We cannot fulfil our human potential, or strive to improve our
characters, our relationships and our awareness, without engaging
in activity that inevitably damages other organisms. Every
organism draws its continued life from interactions with other
organisms; what we need from an environmental ethic is an
understanding of the limits to be observed in our own interference
with other organisms, and with the abiotic environment upon
which we all depend.

Further, we also need to be able to take account of the fact that
whole systems of organisms may strike us as being more significant than individual creatures, and likewise for species. Rolston
who, in an early paper maintained that we could be drawn to an
ecological ethic by a genuine *love* of nature, suggests principles
along just such lines. Here are some of the maxims he articulates
in his essay 'Just Environmental Business' (Rolston 1986):

Respect life, the more sentient more than the less
Lost individuals can be replaced but loss of species is irreversible: so respect life, the species more than the individual
Respect ecosystems, as proven, efficient economies
Maximise natural kinds
Give care to landforms, fauna and flora which are the home
in which life is set

For someone who takes systemic value as motivating moral commitments, this is an intelligible list. Yet comparing Rolston with Taylor, we have to confront some peculiarities in our intuitions. Although the injunction to respect species more than the individual sounds fine, it is hard to state a rationale for preserving a species of insects in preference to an individual lion (supposing we had the choice). There are, after all, more species of insect than are ever likely to be classified, and there are very few lions. The trouble is that Taylor's account captures some of our intuitions: we do sometimes feel about things in a way consistent with his principles. But Rolston's does too. Even after making due allowance for the contexts in which our intuitions are supposed to work, we are likely to face the same dilemma. Within the confines of environmental ethics we are confronted with the need to recognise that no one systematisation of ethics will do justice to its complex structure.

13.2 Nature, role and function

Consider the destruction of trees to make paper. This paper is then used, let us suppose, in the production of advertisements which will add to the world total of 'junk mail'. Although such use of paper accounts for huge amounts of it, even larger quantities go into the production of tabloid newspapers which, in the UK at least, provide as much misinformation as news. Often, such newspapers fail to provide more than a minimal amount of light entertainment to their readers, and at the worst they can be accused of pandering to the worst prejudices of racism and sexism. When we take stock of this use of trees, it is hard to avoid thinking in terms of something like Taylor's proportionality principle. If things of grandeur, dignity and beauty are to be sacrificed, let it be for reasons of some dignity; let their deaths serve some serious human interest instead of the merely trivial.

But what would make someone dismiss advertising circulars, and the tabloid press, as catering to merely trivial interests? There is, after all, big money in advertising and newspapers. Here I run the risk of sounding elitist. The reason these things seem unworthy beneficiaries of the cost to nature involved, I would suggest, is that they are not things which contribute to the development of human potential in a constructive, educational or enriching way. If we take seriously the view of human beings suggested by ecological analogies, we will look for conditions in society that will enable humans to realise their potential in multiply satisfying ways. But the notion of *satisfaction* here needs to be understood equally ecologically. If many of our significant properties are context-dependent, then our satisfactions will involve interactions with our context of the appropriate kind.

There are things for which the story of how they got where they are — the story of their origin — together with an account of where they are exhausts everything that is to be said about their natures. Elsewhere, I called such things 'intrinsically functional' (Brennan 1984). There are no significant properties of hearts, lungs, feet and eyes that are not described in the account of their nature as functional components within larger bodies. By contrast, the items that we think of (quite properly) as whole *natural objects* have no intrinsic functions similar to those possessed by hearts, eyes and other organs. The funny thing about the possession of ecologically supervenient properties is that they seem to be important, while yet — as seen in Chapter 8 — they are not definitive of the items in question. It would be strange to find a crow or a magpie that did not scavenge in circumstances where scavenging is possible. This deviant behaviour, however, would not make the bird in question a doubtful member of its kind. Yet we know full well that, in the right environments, crows will clean carcasses. We can thus come to regard feeding on carcasses as part of the *function* of crows, their special part in recycling of matter typical of ecosystems and biotic communities.

If things were what they ate, then the story about the natures of things would be less complicated than it is. If things were entirely separate individuals born into worlds in which their fundamental and realised niches coincided, then the story about their natures would also be less complicated. But many organisms are ecological opportunists, exploiting what resources they can and switching their feeding preferences according to the resources available and to the competition they face. Adaptation to, and survival in, varied contexts is the key to speciation, as biologists have recognised since Darwin. A species population that survives in a rather different environment from its predecessors and which continues to flourish despite the problems of adaptation may be the forebears of a new variety of their kind, and hence, ultimately, of a new species. What we regard as the essential natures of the natural kinds around us is an approximation to a description true at a moment of the state reached by a continuous process of change. Adaptation, variation and speciation ensure that many of the species we now think of as fixed and unchanging will simply be staging posts in the development of a genetic line over time. Each individual thing plays its unique part in this process, coping as best it can with the vicissitudes of the world around it, surviving or dying, leaving descendants or not.

So it is not at all strange that we think of individual things in different ways at the same time. Each thing continues a germ line, defined as to its nature by the DNA it carries, and thus by the reproductive opportunities open to it. But each thing that survives for any length of time is a triumph of adaptation, an accumulation of historical successes whereby its in-built programming, its in-built capacities, have been tested and have brought it through successfully. Each thing is thus a complex of historic contingencies and of general features. As argued in Chapter 8, we may also identify things with the species populations to which they belong. The population of crows in a certain locale see to it that no rabbit carcasses are left for long without being consumed. To the extent that we identify a particular crow with that population it is not just a carrier of the genes of its kind, a parent of so many offspring, but also a consumer of rabbits, and a recycler of nutrients.

Switch the focus now and think of humans, located as they are in groups, societies and cultures. Some of the properties discussed by sociologists — class, status, power — are held by individuals only because they are located in certain kinds of social structure. Yet persons as we know them are not merely social constructs, not items to be defined purely in terms of their functions or roles in society. Pettit has called the doctrine of significant, non-individual properties the doctrine of *social holism*, and it is closely associated with the *ethical holism* which I endorse (Pettit 1986). Persons are autonomous. Adoption of social roles, forming relationships, caring about things other than oneself, and pursuing goals in common with others are not behaviours that compromise such autonomy. In accepting life within society, the agent accepts the need to identify with at least some of the currents, movements and groups within that society. A human life, a social life, requires some such identification. Autonomy under the aspect of social holism then makes sense of some of the things we say and do. We may judge people by the friends they have (although this is not always a wise policy), by the political commitments they make, and so forth. On Kant's account of autonomy, the paradox of freedom is resolved by the claim that autonomous agents have the power of will, but as rational beings they will what is good. The free will submits freely to injunctions issuing from its own rationality, and is thus self-determining.

The social extension of all this is that I accept certain identifications as constitutive of who I am. I have a certain freedom to choose my identifications — to choose to support this or that party, to marry this or that person, to care about this or that cause. We have already noted in passing the possibility that our freedoms may be less than we think — because of our biological natures, or because of the power of our unconscious desires, or whatever. I have not, however, been pessimistic about the likelihood of freedom, even while recognising that it is not as great as pure moralists might like to think. But a life of growth, enrichment and continuing worthwhile experience requires some identification on the part of the subject with items outside the subject. Love is associated with one form of identification, although not all identification involves love. And weak forms of identification allow that I may be unable to act in any way towards the good of those groups with which I do identify. I can still feel pride or outrage, shame or joy in respect of the fortunes of groups to which I do not belong, although our primary identifications, no doubt, are with groups to which we do belong.

The phenomenon of identification, of coming to care about things other than ourselves, and of seeing ourselves as committed to their good, welfare or preservation is one whose recognition provides the possibility of a solution to two problems. First, there is the issue of self-realisation. It is clear that self-realisation is not a project that humans may undertake in isolation from other things, although it may involve periods of isolation (for example, from other humans). Just as Dewey suggested that individuality is manifested through participation in communal, group projects, so self-realisation is typically achieved through life in one or more communities whose ideals, projects and commitments are definitive, to an extent, of the projects, ideals and commitments of the individual. Second, we now have the seeds of a solution to the problem that Mill failed to solve, the problem of the communal and individual good.

As long as the individual is taken as separate from, and aspiring to goals incompatible with the community, Mill's problem is bound to arise. For, as the tragedy of the commons illustrates, there is every likelihood that pursuit of individual goods will be incompatible with maintaining the common good. But now think of the individual as constituted, in some sense, by commitments and identifications that are internal to the community, and the community as rich enough in what it offers to support a multiplicity of individual routes to self-realisation. There need no longer be any failure of fit between what the individual wants to achieve and what is in the interest of all. For individuals identify the good of all with their own good. Just as my own self-realisation is not independent of the good of my family, my department, my town and my country, so, in general, there need be no trade-off between self-development and general good (a point recognised by both Butler and Berkeley in the eighteenth century). The two come into conflict only when individuals lose their ability to identify with the groups around them, lose their place, so to speak, in society and come to see themselves as being fundamentally at odds with the general good of the groups within which they are located. This condition is very similar to what Marxists regard as a state of *alienation* (Naess 1984, where 'alienation' is defined as the opposite of 'identification').

Ethical or moral holism, then, involves a perspective on human nature. It takes seriously the idea that humans are social beings, finding their fulfilment in social living. Human beings are autonomous, of course, if we mean by this that they are, at least, lacking in intrinsic functions. Who they are is then to some extent a matter of the commitments they take, the groups to which they attach themselves, the identifications they make. The prospects for their own self-realisation are strongly affected by such choices. In an extreme case, where individuals can no longer identify with the society around them, or with suitable groups within it, we have a situation of alienation, and thus a situation where possibilities of fulfilment are denied. Human nature is not, let us remember, a matter purely of biological or social determinants. Unlike the crow, humans have a choice over which identifications they will make, over which supervenient properties they will adopt. Put a crow in the right ecosystem and it will soon be devouring a carcass. But put human beings in a social context and different people will make different choices, taking on different roles, cares and responsibilities.

Among other possibilities, we have the option of choosing to identify, however weakly or strongly, with other natural things, with systems of them, and even with the land forms which are in Rolston's phrase 'the home in which life is set'. Failure to make such identification is a symptom of alienation from nature. It may be that in the past, or in some of the world's economically poorer societies, such identification is a matter of course. It has certainly been suggested that the indigenous culture of the original American peoples was one which involved a serious identification with tribe and with nature (Callicott 1983). If this is true, then my own view is that, for all the wealth we have in the rich countries, we are immeasurably poorer in spirit than those for whom identification with nature was more a matter of course than of deliberate choice. In terms of ecological humanism, our alienation from nature is also a kind of alienation from ourselves, a failure to recognise ourselves in our real location in the world.

13.3 Taking stock

What has been argued so far is that any ethic by which we are to live has to recognise our location in natural and social systems, and take account of our place in history. The story of just what we and other living things are is a complex one, involving properties we possess in our own right but also involving distinctive interactions between ourselves and our surroundings. To some extent we, and other living things, manifest types of features, types of behaviour and so on which, in their combinations, are distinctive of us, while no given kind of behaviour or response is itself unique. An important part of the story of what a thing is concerns where it is and how it got there. For things that are defined by their intrinsic functions — organs, and artefacts (what Lotka called 'exosomatic organs') — where they are and how they got there exhaust the story of their natures. For other things, though, such an account is only partial; their autonomy, their ability to forge an identity amidst the vicissitudes of their environment make them something more than merely functional components of larger systems.

The difference between humans and other living things,

however, is one of degree, rather than of kind, as far as freedom is concerned. Each organism makes choices and decisions under constraints and our human choices are made with more information than are the choices of crows. But just as a living thing has some significant properties due to its context, so even the simply inanimate forge their identities and have characteristics that reflect their environment. A living thing maintains its identity by surviving the challenges of competitors, the hazards of its environment, adopting appropriate modes of co-operation and using energy and matter drawn from its surroundings to maintain its physical structures. But a land form, a rocky crest, a river, likewise persists in the face of pressures of drought, erosion and the rock cycle. Some of these pressures act over such long time scales that we, with our limited view, are virtually blind to them. But a mountain carries the marks of its environment and those species which have associated with it over time just as surely as we carry the marks of our own particular histories and contexts. Recognition of all this in no way supports the doctrines of metaphysical holism dismissed in the first part of the book. Rather, what it indicates is the common situation of all things on the planet, whether living or not. The ecological history of a mountain range involves an account of the species which have over time grown on it, grazed it and quarried it. We can leave our marks on mountains and a chalk cliff is little more than a pile of skeletons. But such things likewise leave their marks on us: as challenges to climbers, providers of desolation to ramblers, objects of beauty and awe to observers. They, like us, just are, in one sense. But in another sense, they become what they are through the interactions they have had with the items that surround them.

It should now be clear why I prefer to resist using notions like 'good', 'worth' or 'value' in application to items like mountains, cliffs, rivers and other natural features. What they share with us and all other intrinsically functionless things is the fact that they can be viewed simultaneously under the two aspects already discussed — the aspect of the independent, particular, highly individual thing, and the aspect of the thing which takes on from its surroundings features of some significance. The double aspect can be described in various ways. We can talk about the intrinsically functionless thing taking on, or having assigned to it, roles or functions. We can talk, in the case of human beings, and some other animals, of identification with other things which does not compromise the autonomy of the individual thing — but is, indeed, an aspect of that autonomy. We can use the notion, likewise, of an item realising its potential, when again we think of it doing so in the actual circumstances within which it exists. However, if we do want to use terms like 'worth' and 'value', then my position is that there is only a difference of degree, not of kind, in the worth, or value, of a psychologically complex creature and the worth of something psychologically simple or not a possessor of any psychology at all. However, my position can be defended without appeals to the existence of natural or objective values.

Up to a point, this puts my version of environmental ethics back into alignment with those of writers like Rolston, Taylor and Attfield. Attfield's position, however, seems to me to be crippled by his commitment to utilitarianism, whose defects I have already elaborated. The extension of moral standing to all natural objects also involves a serious extension of the biocentric outlook shared by both Attfield and Taylor. I am not denying the importance of respect for life in our ethical thinking. Rather, I am claiming that there is a perfectly clear strand in our thinking about nature that involves respect towards all sorts of non-living things. In fact, it is Rolston's view which, as a modern exposition of Leopold's land ethic, comes closest to my own. This is not because I have a commitment to systemic value. Rather, natural communities and systems, in my view, are the home in which life is set, leave their mark on the various lives lived within them and are marked in turn by those lives. Such communities are coalitions of individual things, each of which is the thing it is in part by being where it is, and which, in itself, has no functional role to define it. From these considerations ethical holism derives. Objects, systems, even the land forms around me deserve my respect, deserve ethical consideration simply by being what they are, where they are and interacting with other items in the way they do. Of course, recognising the moral claims of natural things in this way is still to leave open the fact of human intervention governed by priority principles that are an extension of the sort proposed by Taylor. (The general form of one such ethic is given in Regan 1981.)

If we follow Regan's account, we can consider an escalation along the following lines. First, we accept that all natural objects, communities and systems are proper objects of moral considera-tion, based on the grounds suggested above. Regan put this point in the vocabulary of 'inherent value', but — as already suggested — it is possible to give an apposite reinterpretation of this notion in my preferred terms. I would claim that the status of maintaining identity while interacting with its environment gives a natural object a certain dignity, and makes appropriate attitudes towards it like admiration, awe and respect. Finally, Regan moves from the recognition of the appropriateness of such attitudes to the articulation of a *preservation principle*, 'a principle of non-destruction, non-interference and, generally, non-meddling' (Regan 1981: 31). To take such a principle seriously is to take seriously the view that the prospect of human benefits alone does not justify overriding the preservation principle. Only when the human benefits are of the right kind, and there is no other option readily available, should we give consideration to overriding the preservation principle. I cannot accept Regan's preservation principle as he gives it, for on my account we have to recognise that intervention in natural systems is an inevitable aspect of the human situation. But there is an important ethical claim to be made which is closely related to Regan's. This is that our intervention in nature should be gentle and preservationist, not destructive and harsh. This can be maintained as the ecological humanist's preservation principle and — like Regan's one — is likely to conflict on occasion with other ethical principles.

Since various of the communities we inhabit leave their mark on us, real ethical choices involve problems arising for us as members of these communities. But the notion of community — like the notion of ecosystem — is a theoretical one, and communities can be defined at various levels of organisation. Some of the choices we face concern us as members of global communities, participating in the great cycles of matter that are so essential to sustaining life on earth. But other choices confront us as members of much more local communities, groups of only a few members, or social groups with clear administrative boundaries — towns, states, provinces or nations. Many of these choices are not directly matters of ethics, but the approaches we take to them are likely to be different if our thinking is informed by ecological considerations. Hence, an ecological ethic which motivates respect for both individual natural objects and systems of individual objects will deliver immediate practical consequences. Some of these are sketched in the following chapter.

In unpublished work, Peter Wenz has attacked my characterisation of the grounds of moral respect by suggesting that there is no principled way of counting the natural objects in the world, and that there is nothing very odd about the notion of natural good-for-nothings. The first objection can be illustrated by thinking about the difficulty of determining how many objects constitute the Grand Canyon — one, several thousand geological formations, or countless billions of rocks and minerals. There is indeed a difficulty here, although I am not sure of its importance. Some of our geological and political identifications of objects are artificial or at least theory-laden; such artificial entities (for example, continents, mountain ranges, and so on) are constituted of natural things, but not always in unique ways. Thus a mountain range may consist of various strata (themselves identifiable natural objects within geological theory) and of various mountains and valleys (also natural objects). What we call one forest may straddle a number of distinct climatic zones and form many different ecological communities. The general, metaphysical problem of what an *object* is can easily be fuelled by such examples, but — for practical and ethical purposes — what matters most is that readers can form a conception of the class of items for which moral standing is being claimed.

Wenz's second objection, however, has to be dismissed as ill-conceived. Since natural objects, on my account, have no defining or intrinsic functions, there cannot be natural good-for-nothings, or — if anything is a natural good-for-nothing — every natural thing is a good-for-nothing! There may be natural objects to which we owe minimal respect, respect so trifling that we would not seriously consider making allowance for them in any circumstances. But no natural object is defined by its being good-for-something, for although we, and all other living things use — and are used by — the items around us, such use is not definitive of us or other things.

A third objection put forward by Wenz is that being of natural origin does not make something an end in itself; for even an artefact can come to be valued as an end in itself. I agree that artefacts can come to be so *valued* — indeed, Lewis's theory of intrinsic and inherent value fits our attitude to art works rather well. But the point about moral considerability of natural things, although it is meant to motivate respect and admiration, is independent of any actual valuations made. We might live in a world in which no-one valued natural things. On my account, such things would still have in common with us their origin, would still be describable in the ecological, or double-aspect, terms I have suggested, and would still, therefore, merit ethical consideration — even if none were in fact extended to them. In the sense in which Regan and Rolston use the notion, 'value' must capture some objective feature of the world. We live, as they put it, in a world of value, within which our human activities have the possibility of value. However, it is better, I think, to avoid such a controversial use of the notion of value and stick instead to the natural origins of things and their ways of maintaining their identity as grounds for moral consideration of them. If we extend such consideration to them, we reveal ourselves as valuing them, and this valuation is grounded in (though not deducible from) their possession of the characteristics I have suggested.

14

Practical Matters

14.1 Political and economic frameworks

In this chapter, I consider a few of the practical ramifications of the adoption of the ecological perspective urged here. A theoretical work like this is hardly the place for spelling out detailed policy recommendations, not least because the same underlying approach can lead to divergent views about the right policy to pursue. Moreover, the fact that ethics is polymorphic means that we will have to balance very different kinds of ethical considerations against each other in any situation of serious deliberation. Thus respect for persons, respect for life and respect for natural, non-living things can foster attitudes and feelings which will on occasion come into conflict and make real decision difficult. Some suggestions about possible lines along which policy can be developed will simply give content to the abstract reflections engaged so far, and perhaps remind us that philosophy and ethical theory are not so far removed from issues of everyday decision as moral philosophers are sometimes apt to pretend.

We live in a world where our economic thinking seems to be out of touch with what is known about ecology. Instead of designing production processes whose output can feed other processes, we often rest content with a high level of waste produced at a relatively early stage in industrial cycles. By contrast, ecological models show clearly that the detritus of one organism is a resource for another, and 'waste' from one whole system is in turn a resource for another. Of course, all living processes ultimately produce material so disordered that it is no longer a resource to any living things — and thus we obtain the final, high entropic waste of living systems. Energy, as opposed to matter, does not, properly speaking, cycle through ecosystems at all, for energy cannot be re-used, even though it can be locked up for a time in matter passing through a system or number of systems. But although energy is lost at all stages of living processes, it is we humans who seem to be particularly ingenious in harnessing it while being adept at using it inefficiently.

Human industry is not only a massive producer of vast quantities of waste matter and waste heat, but these wastes are often dealt with in an irrational manner. Sewage of various sorts is regularly treated not as a potentially valuable resource but as something to be dumped on those parts of the environment least able to handle it. Rivers, and the biotic communities they support, can easily be destroyed by such treatment, and are additionally used as conduits for toxic chemical material. Air pollution causes billions of dollars worth of damage on a global scale every year, revealing that the tragedy of the commons can apply to the pollution of a common resource as much as to the consumption of one. In the case of air and water pollution, individual or corporate polluters often contribute only in minor ways to a disaster that affects all living things; without fairly imposed control by some form of central agency, no individual or corporation has a motive to clean up their own act while others reap economic benefit by staying dirty.

Our judgement of the good sense and morality of industrial and other practices that impinge in these ways on the environment is overdetermined. From a human-centred perspective, it is possible to deplore the handing down of problems to our children and their children; it is also possible to deplore, from the same perspective, the damage done to buildings by air pollution, the aesthetic offence caused by dying woods and dead lakes, the harm to sporting and recreational pursuits, and the health hazards caused by the introduction of toxic substances to an environment ill equipped to break down and render them speedily available as resources within ecosystems. From the perspective urged in this book, and that taken by Rolston and Regan, it is also possible to deplore environmental degradation as a symptom of insufficient respect to things of natural origin. We have here a situation where we owe respect regarding the environment, and also owe respect — if my suggestions are right — to the environment. To treat it in the way we currently do is also to show scant consideration for ourselves, given the ecological holism which I have suggested motivates preservationist attitudes.

one possible move, at local and national level, is to set up environmental control agencies that are able to enforce standards for the common good, or to prevent the unintentional destruction of items worth respect through millions of acts that are individually trivial. The actual operation of environmental protection agencies is not usually seen as the implementation of specific ethical policies. Yet it is clear that, as soon as we try to deal with actions of many agents all at once, we need more than ethical principles to be brought into play. An accumulation of acts which are either minor wrongs in their own right, or not even wrongs at all, can lead to a state of affairs that contravenes the preservation principle. In short, our *ethical* principles supply a motive for *political* action.

This is perhaps not the place to discuss the fine points of difference between ethical and political frameworks. There is, of course, some difficulty about separating a political view from an ethical one: for politics seem to be ethics focused on one kind of problem, namely the problem of how groups of people should conduct their affairs. Thus conceived, politics is not clearly separate from international relations, for if the groups of people are large enough, then the affairs to be conducted involve more than those within nation-states. Nor, on the other hand, is politics separate from the business of smaller groups, institutions and organisations of the sort that are found within states and more local forms of government. What falls properly within the scope of politics, thought of in this broad way, is precisely a concern with the communal good as opposed to the good of individuals. Groups with political aims are precisely those with programmes and principles concerned with how the affairs of larger groups ought to be managed. In saying this, I mean to leave open the possibility of a political claim to the effect that there is no more to the communal good than the aggregate of individual goods. From what has already been argued, it is clear that I would not wish to support any such political doctrine — but it is a political doctrine for all that.

In many of the rich countries of the world, the current orthodoxy about how the affairs of the state should be conducted is heavily influenced by economic considerations. We can think of the matter again in terms of frameworks. The kind of framework supplied by economics gives us a view of people as factors of production and consumption. In arguments about the virtues of free market economics versus centrally planned ones, the economic impacts of alternative policies are measured by reference to the employment, income and consumption patterns of people conceived in the way just mentioned. There is, after all, no other way of thinking about people while staying within those frameworks. But it is precisely this approach that is criticised by those, like Marxists, who maintain that a quite different description of economic activity is needed. For their alternative economic description involves expanding the scope of classical economic theory to embrace ethical considerations. Marxist categories, like that of 'surplus value', involve an implicit valuation of capitalist economics. More precisely, the description of capitalism by Marxist categories sets the scene for a denunciation of a system which essentially involves exploitation and alienation.

This mixing of frameworks provides for a more sophisticated account of social and political activities. But of course it still leaves the account essentially incomplete, incomplete in the way that all accounts of human activity will be. For any framework — even a scheme that results from the merging of a number of other frameworks — involves limitations, namely the limitation of seeing the subject matter in terms of that framework or scheme and not in terms of any other. At the level of complex phenomena, we have multiple choices among descriptive and explanatory frameworks, as noted in the first chapter. What I want to explore here is the suggestion that ethical perspectives can determine to some extent the limitations of other frameworks, or be mixed with them — as in the example of Marxism — to yield interesting constraints on the theories we adopt within those frameworks.

Clearly, taking preservation seriously means at the very least sustaining the natural world and thus engaging in social, economic and productive activity that shows respect both for ourselves and for the systems within which we are nested. Indeed, we show self-respect partly by showing respect for natural communities as well as social ones. In our economic thinking, we can show a commitment to preservation by opting for steady-state, or no-growth economies where this is appropriate. To avoid confusion, it is worth noting that a steady-state economy need not involve zero growth in GNP, or constant technology, or no change in distribution of wealth within society. As Daly argues, all that is required is an economy meeting the following four requirements: first, a relatively constant population; second, a relatively constant stock of physical wealth (buildings, and other artefacts); third, a balance of humans and artefacts that satisfies sensible requirements on quality of life; and, finally, maintaining the two populations (people and physical wealth) by the most frugal possible throughput of matter and energy (Daly 1984).

Daly's constraints permit change in population, artefacts and wealth. Human populations are inevitably subject to regular variation and replacement as a fact of life. So stability in economics can be thought of as a dynamic equilibrium, not a hold at some current level of technology and physical possessions. What we should try to limit is the maximum number of human beings or the maximum amount of total physical goods in existence at a time. To take this seriously involves an end to the throwaway society, the foregoing of the trivial pleasures of excessive consumption and a recognition of the right of communities to put limits on the fertility of their members.

Outside individual communities and nation-states, a commitment to no-growth economic thinking in the sense just defined would mean taking steps to limit the power of trans-national corporations to operate in a global market. Since global physical wealth would be frozen at a stable and sustainable level, luxuries like intensive research into new weapons systems would have to be given up, and the stock of war materials available to individual states would change very slowly. It is hard to see that this would be a disadvantage, or in any way unethical. However, the control of such matters on an international basis would not be easy.

Perhaps the most ethically controversial aspect of zero-growth economic thinking in the sense defined is the idea that individuals should not be permitted to reproduce as a matter of right. The issue of rights is a complicated one; for, in general, a right claimed by one party involves a cost or other diminishment of liberty to some other agent. Thus my right to walk the streets in safety diminishes the freedoms of would be murderers and muggers and incurs a general cost to society to ensure adequate policing in protection of the right. We might start, then, by supposing that there is a right to reproduce and asking at what cost it is bought. If a male claims such a right, then it is clearly bought at some cost to the woman who is to bear the child. For, she will have to make an investment in pregnancy, and in many societies will be expected to bring up the child. Even if bringing up the child is not, however, her responsibility, it will have to be the responsibility of someone, or of some agency. In either case, there will be a diminishment of someone's freedom, including, in many cases, the burden of tax borne by those who support the necessary welfare services.

But, irrespective of costs, is it not a basic human right to reproduce — just as it is a basic right to have food, to have shelter, and so on? I am not sure just what force the term 'basic' has in this question. It might be plausible to argue that we have a right to impose on our community in respect of those things that are essential to our continued survival in the dignity appropriate to human beings. But it is not clear that the freedom to have children, any more than the freedom to smoke in public places, drive cars with dirty exhausts, or emit noxious wastes from a factory, are ones that society must recognise as essential to dignified human survival. On the contrary, lack of control over these matters is liable to lead to a diminishment of human dignity.

There are big issues here upon which I have not touched. One common theme is the notion that since human life is good, more human life means more good. This claim is popular with utilitarians, and Attfield argues that there is an extensive obligation to bring into existence people who, as far as can be foreseen, will lead happy or worthwhile lives (Attfield 1983: 132). In his view, the only reason for not going forth and multiplying for all we are worth is that to do so would most certainly result in bringing into existence persons whose basic needs would not be satisfied. I have to confess a certain bewilderment when faced with this sort of claim. Where real people are concerned, I agree with the utilitarian that in many cases an incident that kills lots of people is worse, morally speaking, than one that kills only a few. This does not, however, lead me to believe, where merely possible persons are concerned, that a world with five billion happy people is better than one with four billion happy people.

Children yet to be conceived have no right to life, as far as I can see, even though we can talk of the rights of future generations, who are as yet unconceived. Further, it makes perfectly good sense to say that someone has no right to exist, indeed, should not have existed, but to recognise that, since the person now does exist, we have duties owed directly to that person (Parfit 1984: §122). Lastly, of course, there are various religious arguments to the effect that contraception is immoral. To discuss these fully is beyond the scope of this work; but it is worth noting that such arguments seek to impose duties on, and hence diminish the liberty of, human beings in respect of their own bodies and bodily functions. The community that restricts its members' freedom to reproduce at least requires less interference than this. For what is being restricted is not natural bodily activities but foreseeable consequences of such activities.

To say all this is not to solve the problem of how a democratic, liberal society can hope to limit its members' fertility in ways that respect human autonomy, human control over their bodies and recognise that many people would find their lives impoverished if unable to devote part of them to the rearing of children. The satisfaction of having, and bringing up, children is of enough significance to merit priority over respect for other natural things — up to a point. But once a sustainable size for a community has been found, the difficulty is so to arrange matters that the birth rate yields the right replacement while individual freedom is maximised. One obvious economic device fails this test. If having more than a certain number of children draws economic penalties of a severe enough kind, then only the wealthy will be able to afford the number of children they want to have — contrary to the ideal of justice. As a fairer alternative to this, some economists have proposed distributing 'birth licences', or licence units, on a strictly equal basis to individuals within the community, but then allowing a free market in the voluntary exchange of these among people. This proposal has the virtue of allowing control at the level of the society as a whole combined with freedom at the micro-level.

Daly suggests that, in addition to these features dealing with fertility, any society committed to the steady state ought to reduce the inequality endemic in all the industrialised societies by the expedient of setting a guaranteed minimum income and by establishing additionally a maximum on wealth and income (Tobias 1984: 99). Such a proposal recognises one of the most glaring failures of economies that are committed to growth. This is that, contrary to the reassurances of those who argue in favour of growth, the normal increase in GNP associated with successful economic growth does not encourage the 'trickling down' of wealth from the rich to the poor. Instead, countries like the United Kingdom have shown, during their most sustained periods of economic 'success', an ever-widening gap between the rich and the poor and an increasing concentration of wealth and property in fewer and fewer hands, Porrit 1984. It is hard to believe that it is merely a coincidence that a system that is so manifestly unfair to those who live under it also happens to have none of the features of sustainability and operates with precious little respect for natural objects and systems.

I do not expect every reader to agree with the recommendations just mooted. However, they indicate one way in which political and economic thinking can be influenced by an ecological turn in ethics. For those who think that due respect for nature is compatible with the normal targets of economic growth my suggestions will seem misguided. They, however, owe us some account of how they see the destruction of nature, the pollution of air, waters and land, being brought to a halt using policies deriving from conventional political and economic frameworks. The adoption in the right circumstances of zero-growth economic values would be part of the modification of frameworks which might yield what is sometimes called a *paradigm* within which to think about our relations to each other and to nature (Routley 1983).

To conclude this section let us think about growth in absolute numbers of species. Rolston's advice to 'maximise natural kinds' is stated as a kind of ethical maxim (Rolston 1986: 156). Yet a steady state environment strikes me as no less desirable than one in which the number of species goes on increasing. It is hard to give a plausible estimate of the number of species currently on the earth, partly due to taxonomic wrangles, and partly due to our ignorance of the representativeness of our current sample. The usual estimate is that there are certainly more than five million, and possibly around 20 million. Now, a steady-state environment would not prohibit the development of genetic life-lines and the appearance of new species. It would simply be one in which the grand total stayed within some upper limit. Just as 'the more the merrier' style of utilitarianism baffles me, so it strikes me as odd to prefer a world of expanding species numbers to one where there is a steady state.

Of course, the species problem we face today is not one of expanding numbers or a steady state. 'Unless appropriate management measures are taken over the longer term, at least one quarter, possibly one third and conceivably a still larger share of species existing today could be lost' (WCED 1987: 152). My qualms about the principle of 'the more life, the better' should not be taken to imply that this kind of species loss is something to be accepted with equanimity. As with the case of real, versus merely possible people, the preservation principle, as I understand it, requires our respect and care for the species we currently have.

14.2 International co-operation

Any thorough work on 'green' politics (like Porritt 1984) is likely to address a wide range of practical issues. This book is not a further tract on the green programme, so I will have nothing to say about the details of energy policy, the question of nuclear power, the safety of radioactive wastes, the degree of global warming anticipated over the next few years and the associated problems of sea-level rise, nuclear weaponry, conventional war and its impact on remote wilderness areas, priorities for research and development in a world in which half of technical research expenditure goes on weapons research, and a whole host of other matters which are of immediate concern (an appropriate summary can be found in WCED 1987). Even if our focus were of the narrowest, simplest, crudest anthropocentric kind, it would be clear that there are great things wrong with the world today. Two and a half days of global military expenditure would fund a five year programme to save the tropical forests; ten days of current military spending allocated each year to the UN water and sanitation decade would have given the entire Third World clean water and thus vastly improved health (figures from WCED 1987). In terms of the admittedly vague notion of alienation, it is hard not to see our alienation from nature as a symptom of a widespread alienation of one person from another, one way of life from another and one nation from another.

The advance of technology means that we also possess the means to carry the tragedy of the commons to its logical conclusion. We now, for example, have the technology to follow the great circulatory patterns of the oceans with roughly the same accuracy as meteorology can plot the systems of the air. Plotting these ocean movements also means being able to plot the movements of shoals of fish in the deep seas. There is no reason why we should not suck the seas dry of fish, thus adding one more chapter to the story of how a common resource is pursued to extinction. Yet if we fish the seas to extinction, we will not be able to plead ignorance. Likewise, we have the knowledge to manage the planet in a suitably respectful, caring way, just as we have the skills to destroy much that will not readily be replaced. At the moment, there are certain fashions in managerial style — fashions that seem to infect government, education, industry — in the economically richer countries, fashions that themselves show little respect for the ideals of autonomous personhood. It is as if lack of care for persons, and lack of care for nature, are intimately connected, despite the difficulty, at a theoretical level, of showing comparability between the natural and the personal.

If these remarks make our situation seem bleak, then it is perhaps worth dwelling on the more optimistic side of things. First, a word of only slight comfort. The roots of our environmental crisis go back a long way. There was no 'golden age' in the past when our ancestors showed a suitable respect for nature. Thus some of our environmental problems likewise go back a long way. The death of forests, the pressures on water systems may be things that strike us forcibly at the moment: but the ultimate causes of these disruptions may be found several centuries ago. The project of making restitution, or returning natural systems to a state in which they are moderately sustainable may involve more than returning them to their 1950 or their 1940 conditions. We are witnesses to one instant of a prolonged period of challenge and damage to global life-support systems. What we must then recognise is the need to look carefully at our strategy, identifying urgent cases for first-aid while — more slowly — working out longer term strategies for coping with deep-seated ecological disorders,

If we do not share the whole burden of guilt for the mess our environment is in, we do have an unrivalled opportunity to start doing something to put things right. There is no shortage of suggestions from concerned politicians, economists and biologists. These range from the flippant (pipe industrial pollution directly to the homes of directors of the industries concerned) to the serious. The binge of consumerism that has spread through the rich economies this century has to be recognised for what it is — a shallow and unsatisfying pursuit of what is in the end of little real value. This is not to deny that consumerism has its rewards. There are pleasures a-plenty to be gained in the rich societies, just as there are pleasures to be gained from other obsessive pursuits, like the pursuit of sensual pleasures. None of these features, however, either commends such a life or shows it to be proper to our natures. If a general moral repugnance could be generated against the excesses of such a life, then there is no reason why we should not find people in the rich countries opting for modest consumerism. Think, for a parallel case, of the recognition among US citizens of the wastefulness of big, gas-guzzling cars. My suspicion is that at least some of the movement away from larger cars was promoted not by purely economic considerations but by a genuine desire to adopt less wasteful styles of life.

The future of the world is, if present trends continue, largely an urban one. The ecological dream of returning to more natural modes of life, the idyll of the self-sustaining smallholders with their organic plots, and the solution to unemployment of getting the urban poor back to the land are all dangerous myths. They are dangerous not least because the attempt to act in accordance with them leads rapidly to ecological disaster. Current development of the Amazonian basin shows quite clearly that there is just not the managerial expertise available in that kind of area to bring about the successful relocation of urban poor in sustainable agriculture. Instead, vast tracts of tropical forest are destroyed in pursuit of a solution to economic problems which remain no more tractable after the damage is done (Moran 1984).

Exploitation of environmental resources is all too often regarded as a kind of cheap and quick solution to a problem that is endemic to the political and economic structure of a society. It is essential for governments in countries that still possess vast and impressive natural resources not to buy time in this way. Just as it is no solution to the problem of failing fisheries to keep fishing until the stocks are exhausted, so it is no solution to problems of foreign debt, high unemployment and runaway inflation to squander the last of a nation's natural resources in the hope that something better will turn up tomorrow. Instead, as the experience of sub-Saharan Africa has shown, what happens tomorrow is very likely to be worse than the situation today, and worse because of the very environmental degradation that current policies encourage.

What, then, is the solution to the problem of the urban future in countries whose poverty seems to leave them with few options? What has happened all over the globe is that the rural poor have migrated to the cities in search of opportunities. They find there an environment scarred by uncontrolled industrialisation, with heavy and light industry alike free of the pollution curbs to which they are subject in Europe, Australia and North America. Urban planning is virtually non-existent, sanitation and water arrangements are hazardous to health, and transport chaotic. The standard of even public buildings is very low, and many citizens live in shanties, or other temporary dwellings on sites which they have temporarily won in competition with others. It is hardly surprising, then, that around 60 per cent of the population of a city like Calcutta suffer from respiratory diseases (pneumonia, bronchitis, etc.) and that dysentry, hepatitis and typhoid are epidemic in many urban centres (WCED 1987: Ch. 9). The kind of local government structures that many poor countries have inherited from their colonial past are often quite incompetent to deal with these problems. Nonetheless, with resourcefulness, often cheerfulness, human beings struggle to maintain in these conditions a form of life that is more than mere survival.

In a complex situation of this sort, there are numerous strategies to be tried that are well known to those expert in the field. Sensitive overseas aid can help local administrations improve their ability to cope with the massive difficulties they face. Controls on industrialisation, and the siting of industries, can take the pressure off the cities. However, as in the case of what we are assured are 'safe' nuclear reactors, we have to take care not to site dirty industry away from public view. If keeping industry near the urban centres has any virtues, one will be the pressure that this ought to bring to maintain high safety standards and strict control of pollution.

Human beings are by their very nature highly adaptable. Thus they can cope with rural isolation just as they can cope with over-crowded cities. More than coping, however, the people who live in urban centres have an appreciation of their immediate problems, and — sometimes — an awareness of their remoter causes. Decent overseas aid targeted on such people (and very little of it currently is) should avoid the error of *assistentialism*, to use Freire's ugly term (Freire 1974). It is one thing to conceive a housing aid programme for the poor of Calcutta while sitting in a comfortable library somewhere in Europe or North America. It is quite another thing to offer expertise in a trusting way — a way that trusts the judgement of those on the spot. Put another way, it is easy for experts to forget the limits of their expertise; to regard themselves as the sole people who understand problems and can come up with solutions. But the really useful expert is one who can engage in dialogue with the people who require help, learning from them and enabling them, in turn, to help themselves in the situations in which they have to live. The same might be said about respect for, and learning from, nature — again reminding us of the parallel between respect for persons and respect for nature.

The enabling conception of the expert is not only one relevant to the Third World. It is very much needed in the industrialised societies which all too frequently run local and national government in a paternalistic fashion. Town and city planning all too easily becomes a matter of finding solutions *for* people, not *with* them; consultation becomes a matter more of public relations than of listening to the needs, understanding and ideas of those for whom the plans are being made. The ethical objections to paternalism and to assistentialism do not require anything so arcane as an ecological ethic. They simply require us to bear in mind the respect and dignity which is owed to human beings as free, self-regulating creatures with the capacity for lives and projects of value. It follows almost immediately that we do less than justice to such creatures if we do not involve them in decisions that affect the quality of lives they can live, and the nature of the environment against which their projects are to be developed. Sometimes, it will be objected, people lack the resources to appreciate their real problems and their true situation. Although I am sceptical of objections of this sort, I will look shortly at the prospects for the education of just such people.

The move towards *community architecture* in the UK and elsewhere in Europe is a good example of what happens when paternalism and assistentialism are avoided. Community architects have skills that those requiring their services lack — knowledge of materials, ability to draw up plans and so on. But they recognise that the community they are serving have something which they, as architects, lack, which is a real interest in the environment which is to be created. Here we are using the term 'real' in a sense close to Newman's, and contrasting it with the *notional* interest and understanding that any intelligent being can possess of the circumstances of another (Newman 1870). Likewise, my own understanding of the destruction of tropical forest is notional, rather than real, for I have never seen the bulldozing, burning and erosion of which I have read such graphic accounts. Of course, a notional understanding of many things is good enough for many purposes. But a notional understanding of another creature's circumstances is a poor basis on which to take decisions that affect vital aspects of that creature's well-being. In the case of non-humans, we have no other means of finding out their views on our policies, and adapting our doings to their needs and concerns. But in the case of other humans, we can recognise their authority on such matters by giving them the power to make decisions for themselves on matters that most concern them.

The very process of consulting and engaging in dialogue with those whom we seek to help is itself a potent force with which to liberate their own ideas and thinking about their situation. This is true in the case of Freire's educational programmes, and it also seems to be the case with those community architecture schemes about which I have information. Engaging people in thinking critically about themselves and their own situation is a precondition of taking steps that will lead to an improvement in their situation.

Applied to the urban poor of the Third World, aid schemes that take account of their autonomous, human status would favour establishing and funding neighbourhood organisations which are trusted to see to their own needs. The building of local clinics, improvement of roads, establishing of safe water supplies, and so forth will be the responsibility of the neighbourhood group, assisted by sensitive experts and by the necessary finance. It is sometimes suggested that such schemes are also among the most cost-effective and usefully labour-intensive of the forms of aid that can be given. If so, here is a happy consilience between economics and humanity. Such schemes also have the potential to be educationally valuable, and more will be said about this in the next section.

A practical idea for specifically international co-operation concerns the control of the multinationals and the problems of industrial development. Whereas an individual nation may be powerless to deal with a large international combine, a regional grouping of countries represents considerably more power. It could become common practice to vet all proposals for industrial development within a region, and call for environmental impact assessment of any that are ecologically significant. The control of the transport of waste or hazardous materials in any fashion across national boundaries is also best done at regional level. After all, it is now widely recognised that pollution knows no boundaries, so it is only sensible to have regional co-operation in the monitoring and control of such pollution.

These suggestions show the possibility of combining respect for local participation with a new kind of internationalism (Porritt 1984: Ch. 12). I do not, however, agree with Porritt's suggestion that the future of the planet will involve some kind of world citizenship or that we will be able to move to more rural lifestyles again. For the foreseeable future, I suspect, we will have a continuance of the drift to the cities, and therefore a real need to address the problems faced by city-dwellers. City-dwelling suits our social nature, after all, and arguably preserves more of the natural systems in a state uncontaminated by much human intervention. What matters is that people live on a human scale, while taking the necessary action at an international level to control the identified threats to our lives, and the systems on which they depend. Preservation, as understood by the ecological humanist, means, in political terms, a move towards less exploitation of natural environments, and a containment of the restless power of industry and consumerism.

If the earlier claims about the interrelations between humans and the environment are correct, we should be able to look forward to real benefits accruing to us from a greater respect towards our shared earth. Care for our natural surroundings offers the prospects of transcending local, cultural or national barriers to mutual understanding. Joint projects involving several nations focused on items of common concern are a good basis for building mutual esteem. To start doing good towards nature, would — human psychology being what it is — mean that nature would itself come to mean more to us than it seems to do at the moment. But since nature bears our marks, and we are marked in turn by it, an increase in care for our environment would match an increasing sense of our own dignity as members of natural systems. There is no panacea, no global solution to our problems, in an ecological ethic. For we badly need a social ethic that allows the dispossessed to enter once more into membership of communities that are rent by poverty and hopelessness and which manifest the symptoms of these underlying diseases — violence, crime, social disorder. If we really do think of persons as rational consenters to a social contract, it would be as well to try to reflect this conception in our social structures. In this respect, many of the 'developed', or 'advanced' countries of the world could take a lesson from those which are labelled 'undeveloped', or 'primitive'. But, as will be seen when we turn to the issue of education, it will require a revolution in the thinking of those with power and wealth to learn to adopt a suitably humble, loving and trusting attitude towards those who are poor, uneducated and without power.

14.3 Education and other animals

Ecological humanism (hereafter *eco-humanism* for short) is, like any other general position on ethics and human nature, not open to conclusive proof or disproof. It is not a position that can simply be deduced from the facts of our nature as beings in a world, marked by that world and acting in turn upon it. As an attempt to say something more than the utilitarian or contract theories it has a value base. In terms of that base, it argues that modes of life that are true to the ecological dimension of our nature are likely to be more rewarding and satisfying than those that ignore what we are and where we are. Like deep environmental positions in general it runs the risk of being accused of romanticising the natural and of ignoring the reality of nature red in tooth and claw.

The only test of such a position, however, is to try it and see if human life is much improved by its adoption. Humans not only act on the world but act on it under descriptions that they give to themselves and others. They look about them from a point of view that is coloured by their own past experiences, their reflections, their education and their various emotions, beliefs and instincts. The eco-humanist recognises the twin aspects of value in life — the particular and the general — and sees this as a reflection of the ecological point that what we are is in part a construct of where we are (see Chapter 10.3). To have a community role is to exemplify general characteristics that might also be exemplified by others; to be an individual is to be a focus of characteristics that reflect an idiosyncratic path through the world at a certain time. An ecological stance accepts the importance of both the general and the particular, the community influences and the aggregate of marks from them. Value is thus a dynamic, not a static, matter. It is a play, a dance, a transformation of the one through the impact of the many.

I have tried to identify a certain truth in holism. In metaphysical terms, this truth involves the recognition of supervenient characteristics as significant, as part of the story of what a thing is. The supervenience, however, involves the dynamics of the particular item and its nesting within a larger environment of action. Translated into moral terms, we get social holism and then a kind of large community holism. The construction and reconstruction of experience by humans will involve not only social and natural communities. There is a commonality here between humans, other animals, plants and even inanimate natural things. For all forge, one way or another, their identities through processes of interaction with their surroundings, feeding — either literally or metaphorically — off the things around them while maintaining their own separateness. Eco-humanism has needed no stronger form of holism to support it.

As developed, the position has recognised the likely tensions between socially and environmentally derived moral thinking. These tensions are particularly clear when it comes to our treatment of other animals. Vegetarianism and hunting pose special problems for the eco-humanist. Let us consider hunting first of all. On one perspective, hunting for pleasure can be condemned as a sadistic pastime that degrades the hunter, terrifies the victim and spoils the environment. On another, it builds character, allows us to forge mystical bonds with nature and provides a spur to the protection of habitats that might otherwise be destroyed. Aldo Leopold, after all, came to care about wolves after shooting an old one and arriving 'in time to watch a fierce green fire dying in her eyes' (Leopold 1949: 130–2).

Some of the crudest attempts to justify hunting will certainly not do. Loftin argues that 'pragmatic ethical principles' justify those forms of hunting which, as a matter of fact, result in habitat preservation (Loftin 1984). But all sorts of immoral activities might, as it happens, have no bad environmental consequences or even good ones. Moreover, even if some hunting is acceptable by this criterion, much of what passes for hunting involves severe environmental damage (burning off on grouse moors, for example) or ruthless extermination of normal predators on the game being managed. Since eco-humanists would, I hope, be wary of utilitarian reasoning in any case, these considerations will leave the ethical status of hunting very much undecided.

It would not be unreasonable for different adherents of eco-humanism to come to rather different verdicts on various forms of hunting. As already noticed, subsistence hunting tends to be treated as an exceptional case by theorists like Taylor. Yet it is not hard to imagine other forms of hunting that do involve some real respect for nature. Likewise, there are forms of activities like poaching which are celebrated in folklore as an assertion of the individual against those who are overprivileged and overfed. The solitary poacher on a Highland estate will not be regarded by the eco-humanist in the same light as a gang using machine-guns to massacre herds of elephants in Zimbabwe. Hunting can be done with or without restraint, even when conducted as a sport. The eco-humanist is likely to be concerned with the hunter's motive, the nature of the experience, and the role of that experience in the hunter's life as a whole, before giving any verdict on the morality or otherwise of the activity.

The same, I think, is true of vegetarianism. Meat-eating is an expensive use of land, and many sorts of factory farming degrade both those who practice it and the animals exploited by it. Just as in assessing the ethics of hunting, we have legitimate concern for the terror, suffering and pains of the animals in question. This concern does not derive from ecological considerations, however, but rather from general considerations regarding the infliction of avoidable ills on sentient beings (as argued, for example, in Clark 1984). Eco-humanists can hunt and rear stock in ways compatible with respect for nature, and indeed in ways that enhance their experience of nature and show concern for the sustainability of natural systems. In present circumstances, those attracted to eco-humanism will want to confine their consumption of animal flesh to those creatures which have been reared humanely and whose role has been environmentally enhancing rather than a drain on scarce arable resources. Vegetarianism will thus not be mandatory for eco-humanists, although only modest meat consumption can be justified by the principles sketched.

There is clearly much more that could be said on both of these vexed topics. The vegetarian who is committed to the removal of avoidable suffering to all sentient beings is perhaps not, on that account alone, disqualified from adopting the eco-humanist stance. However, there are bound to be tensions between such a view and some principles of preservation and non-intervention (Callicott 1980). Indeed non-intervention cannot be seen as a real constraint on the eco-humanist. If interactions between us and our surroundings are partly constitutive of our natures, then intervention will be inevitable. The important thing is the quality of that intervention — whether it be modest, respectful and loving, or arrogant, exploitative and unfeeling. Loving, caring intervention in nature will not always be such as to preserve on a local scale, although one sign of a caring attitude to nature will be the desire to preserve some parts of it untouched by much human activity (though for cautionary remarks on preservation versus conservation, see Norton 1986).

Education is traditionally taken to be the hope for the future. Any defence of eco-humanism thus has to suggest ways in which educational practice can be refined so that it can take account of such an ethical stance. I hope that my defence of eco-humanism and of the relevance of ecological frameworks has been undogmatic enough to make it clear that I would not favour large-scale indoctrination. Instead, educational programmes, whether in the rich or the poor countries of the world, and whether aimed at adults or children, need, in my view, to take account of the ideas of two twentieth century theorists — Dewey and Freire. Dewey emphasises what, in Newman's terminology, is the real, as opposed to the merely notional, grasp of issues through inquiry and project work (Dewey 1915, 1916). He never regarded learners in a democratic society as merely passive absorbers of remote information. To treat them as such was to be false to the very ideals of democracy itself.

Something close to Dewey's ideals seems to be recognised in the various declarations of international bodies on environmental education. For example, the 1977 Tbilisi declaration states:

> Environmental education must look outward to the community. It should involve the individual in an active, problem-solving process within the context of specific realities, and it should encourage initiative, a sense of responsibility and commitment to build a better tomorrow.
> By its very nature, environmental education can make a powerful contribution to the renovation of the education process. (Holdgate, Kassas and White 1982: Ch. 15)

The ecological, and eco-humanist, aspects of such a stance make it commendable as an ideal, though far from easy to put into practice. In the rich countries of the world, the overt curriculum of the schools can pay lip-service to such an ideal while the covert curriculum encourages pupils to prepare for a society in which the successful are the strong, the exploitative and those with least concern for community values.

However, a problem arises with adult education in the countries where environmental problems are specially critical. Those involved in the immediate destruction of a common resource often conceptualise what they are doing under various positive labels. They are 'taming the wilderness', making what is 'barren' productive, or bringing order and tidiness to what is unkempt and wild. As already observed, the worst forms of environmental destruction can be checked in the short term by bringing pressure from above. A moratorium on loans for road and railway building in the world's rainforests would immediately check the single greatest source of species loss at present; likewise,

219

a refusal to fund high-technology fishing fleets would temporarily preserve some of the remaining fish stocks.

It would be gratifying, however, if those who are locally involved in the great binges of environmental destruction currently taking place could be reached by education programmes that transformed their image of what they are doing. The challenge can be met in different ways. If we follow Freire's recipe, we will approach the problem of consciousness-raising in a way that respects the learner (Freire 1972, 1974). There are, of course, practical reasons for doing this; but let me dwell on the more elevated, philosophical reasons for such an approach. The learner, as a person, is someone for whom education is meant to be life-enhancing and transforming. But this means that the educator is not just an outside force that impinges on learners. Rather, education comes through interaction between the two sides and the attempt that each makes to learn from the other.

Freire's mode of bringing education, and hence liberation, to the learner involves getting learners to appreciate their historical and political situation. This does not mean political harangues, nor the production by the educator of formulae that capture the nature of the alienation, exploitation or whatever, afflicting the learner. Rather, the educator brings out a realisation of their situation that is submerged within the learners, hidden by 'false consciousness'. In promoting the awakening of the hidden ideas, the educator can learn from the learner, and the learners come to appreciate that they have a voice, something to say. In this way, the 'culture of silence' — what Freire regards as an essential aspect of exploitative societies — is overcome, and a richer conception of their own humanity is made available to learners.

Clearly, Freire's approach to education is politically radical. Yet we need be neither Christians nor Marxists (Freire is both) to learn something from it. Education that recognises, supports and aids the development of personal autonomy will be a powerful force for social and political change. It is easy to blame our lack of action on environmental problems on an educational system that encourages passivity, acceptance and respect for authority (however incompetent). It is clear that education cannot, in a very short time, make much difference to the large problems that confront us. However, with proper educational provision there is every prospect of increasing the number of people around who have a real grasp of the serious problems we face and understand the priorities involved in tackling them. We will certainly know when progress has been made. For this will be shown by the fact that environmental education, in its broadest sense, has come to have as high a priority as literacy and numeracy.

At this point, I rest the case for eco-humanism. These last six chapters have done no more than outline what I take to be the problems with a great deal of modern moral theory. My account of an alternative, humanistic approach to our situation has itself been only schematic. However, I will be content if I have succeeded at least in my negative purpose. Within the ranks of those who care a great deal about environmental issues, a certain position threatens to become almost an orthodoxy. The position challenges current scientific thinking as atomistic and reductionist. It offers holistic alternatives and relates these to a new perspective on health, ethics and life in general (for example, Schwartz 1987). I hope to have shown that this position rests on arguments that make large metaphysical assumptions, that are sometimes confused and that are inessential to supporting the value position of the modern environmentalist.

On the positive side, I have tried to identify what is true about holism in both its metaphysical and moral aspects. Using only modest insights from scientific ecology, I have tried to show that an interesting moral perspective can be developed that denies extreme claims about the plasticity of human nature. On this perspective, an essential aspect of leading an examined and worthwhile life will be taking account of our location in, and effect upon, the natural systems around us. The positive ideas sketched in this work leave open various developments of the eco-humanist perspective, all of which will take account of the ecological setting for human projects. Lives that are worth living will all be lived in nature and the worth that we accord to our natural context cannot ultimately be separated from the worth of such lives.

Bibliography

Achinstein, P. (1983) *The Nature of Explanation*, Oxford University Press, New York.

Aristotle (1955) *Physics* Ross, W.D. trans. Oxford University Press, Oxford.

Armbruster, W. Scott (1986) 'Reproductive Interactions Between Sympatric *Dalechampia* Species: Are Natural Assemblages "Random" or Organized?' *Ecology 67*, 522-33.

Ås, S. (1984) 'To Fly or Not to Fly? Colonization of Baltic Islands by Winged and Wingless Carabid Beetles', *Journal of Biogeography II*, 413-26.

Attfield, R. (1983) *The Ethics of Environmental Concern*, Basil Blackwell, Oxford.

Baier, A. (1986) 'Extending the Limits of Moral Theory', *Journal of Philosophy 83*, 538-40.

Begon, M., Harper, J.L. and Townsend, C.R. (1986) *Ecology: Individuals, Populations and Communities*, Blackwell, Oxford.

Benson, J. (1978) 'Duty and the Beast', *Philosophy 53*, 529-49.

Beven, S., Connor, E.F. and Beven, K. (1984) 'Avian Biogeography in the Amazon Basin and the Biological Model of Diversification', *Journal of Biogeography 11*, 383-99.

Blackburn, S. (1985) *Spreading the Word*, Blackwell, Oxford.

Blackstone, W.T. (ed.) (1974) *Philosophy and Environmental Crisis*, University of Georgia Press, Athens.

Boyd, R. (1980) 'Materialism Without Reductionism: What Physicalism Does Not Entail', in Block, N. (ed.) *Readings in the Philosophy of Psychology I*, Methuen, London.

Brandt Commission (1980) *North-South: A Programme for Survival*, Pan Books, London.

Brandt, R.B. (1979) *A Theory of the Good and the Right*, Oxford University Press, Oxford.

Brennan, A.A. (1984) 'The Moral Standing of Natural Objects', *Environmental Ethics 6*, 35-56.

—— (1986) 'Ecological Theory and Value in Nature', *Philosophical Inquiry 8*, 66-95.

—— (1988) *Conditions of Identity*, Clarendon Press, Oxford.

Brody, B.A. (1972) 'Towards an Aristotelian Theory of Scientific Explanation', *Philosophy of Science 39*, 20-31.

—— (1980) *Identity and Essence*, Princeton University Press, Princeton.

Cain, M.L. (1985) 'Random Search by Herbivorous Insects', *Ecology 66*, 876-88.

Callicott, J.B. (1980) 'Animal Liberation: A Triangular Affair', *Environmental Ethics 2*, 311-38.

—— (1983) 'Traditional American Indian and Traditional Western European Attitudes Towards Nature', in Elliot and Gare.

—— (1985) 'Intrinsic Value, Quantum Theory and Environmental

Ethics', *Environmental Ethics 7*, 257–75.

Cannon, G. (1987) *The Politics of Food*, Century, London.

Capra, F. (1983) *The Turning Point*, Fontana, London.

Chang, D.H.S. and Gauch Jr., H.G. (1986) 'Multivariate Analysis of Plant Communities and Environmental Facts in Ngari, Tibet', *Ecology 67*, 1568–75.

Chisholm, R.M. (1976) *Person and Object*, Allen and Unwin, London.

Churchland, P. (1979) *Scientific Realism and the Plasticity of Mind*, Cambridge University Press, Cambridge.

Clark, S.R.L. (1983) 'Gaia and the Forms of Life', in Elliot and Gare.

———— (1984) *The Moral Status of Animals*, Oxford University Press, Oxford.

Clements, F.E. (1905) *Research Methods in Ecology*, University Publishing Company, Lincoln, Nebraska.

Cody, M.L. and Diamond, J.M. (eds) (1975) *Ecology and Evolution of Communities*, Harvard University Press, Cambridge, Mass.

Connor, E.F. and Simberloff, D. (1979) 'The Assembly of Species Communities: Chance or Competition?, *Ecology 60*, 1132–40.

Corbett, J. (1960) *Maneaters of Kumaon*, Oxford University Press, Oxford.

Daly, H.E. (1984) 'Economics and Sustainability', in Tobias.

Darwin, C. (1859) *The Origin of Species*, Murray, London.

Davidson, D. (1982) *Essays on Actions and Events*, Clarendon Press, Oxford.

Dawkins, R. (1976) *The Selfish Gene*, Oxford University Press, Oxford.

Dennett, D. (1984) *Elbow Room*, Clarendon Press, Oxford.

Dewey, J. (1915) *The School and Society*, University of Chicago Press, Chicago.

———— (1916) *Democracy and Education*, Macmillan, New York.

Diggs, B.J. (1981) 'A Contractarian View of Respect for Persons', *American Philosophical Quarterly 18*, 273–83.

Duhem, P. (1914) *The Aim and Structure of Physical Theory* (Wiener P.P. trans. 1974), Atheneum, New York.

Dummett, M. (1978) 'Realism', in *Truth and Other Enigmas*, Duckworth, London.

Elliot, R. and Gare, A. (1983) *Environmental Philosophy*, Open University Press, Milton Keynes.

Elton, C. (1927) *Animal Ecology*, Macmillan, New York.

Elton, C.S. (1946) 'Competition and the Structure of Ecological Communities', *Journal of Animal Ecology 15*, 54–68.

Emlen, J.M. (1984) *Population Biology*, Macmillan, New York.

Feinberg, J. (1974) 'The Rights of Animals and Unborn Generations', in Blackstone, W.T. (ed.) *Philosophy and Environmental Crisis*, University of Georgia Press, Athens.

Forbes, S.A. (1880) 'On Some Interactions of Organisms', *Bulletin Illinois State Laboratory of Natural History 1*, 3–17.

Fox, W. (1984) 'Deep Ecology: A New Philosophy of Our Time?', *The Ecologist 14*, 194–200.

Frankena, W.K. (1963) *Ethics*, Prentice-Hall, Englewood Cliffs, N.J.

Freedman, D. (1971) *Markov Chains*, Holden Day, San Francisco.

Freire, P. (1972) *The Pedagogy of the Oppressed*, Penguin, Harmondsworth.

Bibliography

——— (1974) *Education: The Practice of Freedom*, Writers and Readers, London.

Galbraith, J.K. (1967) *The New Industrial State*, Penguin, Harmondsworth.

Gause, G.F. (1934) *The Struggle for Existence*, Williams and Wilkins, Baltimore.

George, S. (1977) *How the Other Half Dies*, Allanheld Osmun, Montclair, M.J..

——— (1984) *Ill Fares the Land*, Writers and Readers, London.

Ghiselin, M. (1974) 'A Radical Solution to the Species Problem', *Systematic Zoology 23*, 536–44.

Gibson, J.J. (1979) *The Ecological Approach to Visual Perception*, Houghton Mifflin, Boston.

Gleason, H.A. (1952) 'Delving into the History of American Ecology', *Bulletin of the Ecology Society of America 56*, (1975) 7–10.

Golley, F.B. (1960) 'Energy Dynamics of a Food Chain of an Old-Field Community', *Ecological Monographs 30*, 187–206.

——— (1987) 'Deep Ecology from the Perspective of Environmental Science', *Environmental Ethics 9*, 45–55.

Goodman, N. (1965) *Fact, Fiction and Forecast*, Bobbs Merrill, Indianapolis.

——— (1966) *The Structure of Appearance*, Bobbs Merrill, Indianapolis.

Goodpaster, K. (1978) 'On Being Morally Considerable', *Journal of Philosophy 75*, 308–25.

Grene, M. (1980) 'A Note on Simberloff', *Synthese 43*, 41–5.

Grinnell, J. (1908) 'The Biota of the San Bernardino Mountains', *University of California Publications in Zoology 5*, 1–170.

Hacking, I. (1983) *Representing and Intervening*, Cambridge University Press, Cambridge.

Haeckel, E. (1866) *Generelle Morphologie der Organismen*, Reimer, Berlin.

Hardin, G. (1972) *Exploring New Ethics for Survival*, Viking, New York.

Haskell, E.F. (1940) 'Mathematical Systematisation of "Environment", "Organism" and "Habitat"', *Ecology 21*, 1–16.

Heal, O.W. and MacLean, S.F. (1975) 'Comparative Productivity in Ecosystems – Secondary Productivity', in van Dobben, W.H. and Lowe-McConnell, R.H., (eds), *Unifying Concepts in Ecology*, Junk, The Hague.

Hellman, G.P. and Thompson, F.W. (1975) 'Physicalism: Ontology, Determination and Reduction', *Journal of Philosophy 72*, 551–64.

Hempel, C.G. (1965) *Aspects of Scientific Explanation*, Free Press, New York.

——— (1966) *Philosophy of Natural Science*, Prentice-Hall, Englewood Cliffs, N.J.

Hobbes, Thomas (1651) *Leviathan*, Andrew Crooke, London.

Holdgate, M.W., Kassas, M. and White, G.F. (1982) *The World Environment 1972–82*, United Nations Environment Programme (UNEP), Tycooly, Dublin.

Holling, C.S. (1976) 'Resilience and Stability in Ecosystems', in Jantsch and Waddington.

Holsinger, K. (1984) 'The Nature of Biological Species', *Philosophy of Science 51*, 293–307.

Horn, H.S. (1975) 'Markovian Properties of Forest Succession', in Cody and Diamond.

——— (1976) 'Succession', in May.

Hull, D. (1974) *Philosophy of Biological Science*, Prentice Hall, Englewood Cliffs, N.J.

——— (1976) 'Are Species Really Individuals?', *Systematic Zoology 25*, 174–91.

Hume, D. (1739) *A Treatise of Human Nature*, Selby-Bigge, L.A. (ed.) 1964, Clarendon Press, Oxford.

——— (1758) *An Inquiry Concerning Human Understanding*, Hende, C.W. (ed.) 1955, Bobbs Merrill, Indianapolis.

Hunt, W.M. (1980) 'Are Mere *Things* Morally Considerable?', *Environmental Ethics 2*, 59–65.

Hutchinson, G.E. (1957) 'Concluding Remarks', *Cold Spring Harbor Symposium on Quantitative Biology 22*, 415–27.

Jantsch, E. and Waddington, C.H. (eds) (1976) *Evolution and Consciousness*, Addison-Wesley, Cambridge, Mass.

Järvinen, O. (1982) 'Species to Genus Ratios in Biogeography: A Historical Note', *Journal of Biogeography 9*, 363–70.

Kant, I. (1785) *Groundwork of the Metaphysics of Ethics* (Paton, H.J. trans. 1948) *The Moral Law*, Hutchinson, London.

Katz, C.H. (1985) 'A Nonequilibrium Marine Predator–Prey Interaction', *Ecology 66*, 1426–38.

Katz, E. (1985) 'Organism, Community and the Substitution Problem', *Environmental Ethics 7*, 241–56.

Kemeny, J.G., Laurie Snell, J. and Knapp, A. (1966) *Denumerable Markov Chains*, van Nostrand, Princeton.

Kim. J. (1984) 'Concepts of Supervenience', *Philosophy and Phenomenological Research 2*, 153–76.

Klee, R.L. (1984) 'Micro-determinism and Concepts of Emergence', *Philosophy of Science 51*, 44–63.

Kuhn, T. (1957) *The Copernican Revolution*, Harvard University Press, Cambridge, Mass.

——— (1970) *The Structure of Scientific Revolutions*, University of Chicago Press, Chicago.

Lehmann, S. (1981) 'Do Wildernesses Have Rights?' *Environmental Ethics 3*, 129–46.

Leopold, A. (1949) *A Sand County Almanac*, Oxford University Press, Oxford.

Lewis, C.I. (1946) *An Analysis of Knowledge and Valuation*, Open Court, La Salle, Illinois.

Likens, G.E. (1985) 'An Experimental Approach for the Study of Ecosystems', *Journal of Ecology 73*, 381–96.

Lindeman, R.L. (1942) 'The trophic-dynamic aspect of ecology', *Ecology 23*, 399–418.

Locke, J. (1700) *An Essay Concerning Human Understanding*, fourth edition, Nidditch, P.H. (ed.), 1979, Clarendon Press, Oxford.

Loftin, R.W. (1984) 'The Morality of Hunting', *Environmental Ethics 6*, 241–50.

Lorenz, K. (1941) 'Kants Lehre von Apriorischen im Lichte gegenwartige

Biologie' (trans. in Plotkin, H.C. (ed.), 1982) *Learning, Development and Culture*, Wiley, Chichester.

───── (1966) *On Aggression*, (Latzke trans.), Methuen, London.

Lovelock, J.E. (1979) *Gaia: A New Look at the Earth*, Oxford University Press, Oxford.

Margulis, L. (1981) *Symbiosis in Cell Evolution*, Freeman, San Francisco.

Matthews, G.B. (1978) 'Animals and the Unity of Psychology', *Philosophy 53*, 437–54.

May, R.M. (ed.) (1976) *Theoretical Ecology*, Blackwell, Oxford.

Maynard Smith, J. (1974) *Models in Ecology*, Cambridge University Press, Cambridge.

McFetridge, I.G. (1985) 'Supervenience, Realism, Necessity', *Philosophical Quarterly 35*, 245–58.

McGuinness, K.A. (1984) 'Species–Area Relations of Communities on Intertidal Boulders', *Journal of Biogeography II*, 439–56.

McIntosh, R.P. (1980) 'The Background and Some Current Problems of Theoretical Ecology', in Saarinen.

───── (1985) *The Background of Ecology*, Cambridge University Press, Cambridge.

McLellan, D. (ed.) (1977) *Marx — Selected Writings*, Oxford University Press, Oxford.

───── (1980) *The Thought of Karl Marx*, Macmillan, London.

───── (ed.) (1981) *Marx — The Early Texts*, Oxford University Press, Oxford.

Meadows, D.H., Meadows, D.L., Randers. J. and Behrens III, W.W. (1972) *The Limits to Growth*, New American Library, New York.

Meehl, P. and Sellars, W. (1956) 'The Concept of Emergence', *Minnesota Studies in the Philosophy of Science I*, Minnesota Press, Minneapolis.

Mill, J.S. (1874) 'Nature', in *Three Essays on Religion*, Longmans, Green, Reader and Dyer, London.

───── (1910) *Utilitarianism, Liberty and Representative Government*, Dent, London.

Millar, A. (1988) 'Following Nature', *Philosophical Quarterly*, forthcoming.

Moore, G.E. (1903) *Principia Ethica*, Cambridge University Press, Cambridge.

Moran, E. (1984) 'Current Development Efforts in the Amazon Basin', in Tobias.

Murphy, J.G. (1973) 'Marxism and Retribution', *Philosophy and Public Affairs 2*, 217–43.

Naess, A. (1973) 'The Shallow and the Deep, Long-range Ecology Movement', *Inquiry 16*, 95–100.

───── (1984) 'Identification as a Source of Deep Ecological Attitudes', in Tobias.

───── (1986) 'The Deep Ecological Movement: Some Philosophical Aspects', *Philosophical Inquiry 8*, 10–29.

Nagel, E. (1961) *The Structure of Science*, Routledge and Kegan Paul, London.

Nagel, T. (1986) *The View From Nowhere*, Oxford University Press, Oxford.

von Neumann, J. (1923) 'Zur Einführung der transfinited Zahlen', translated in van Heijenoort, J. (ed.), *From Frege to Gödel*, Harvard University Press, Cambridge, Mass. 1967.

Newman, J. (1870) *Essay in Aid of a Grammar of Assent*, Longmans Green, London.

Niven, B.S. (1982) 'Formalisation of the Basic Concepts of Animal Ecology' *Erkenntnis 17*, 306–20.

Norton, B.G. (1986) 'Conservation and Preservation: A Conceptual Rehabilitation;, *Environmental Ethics 8*, 195–220.

Odum, E.P. (1969) 'The Strategy of Ecosystem Development', *Science 164*, 262–70.

Parfit, D. (1984) *Reasons and Persons*, Clarendon Press, Oxford.

Passmore, J. (1974) *Man's Responsibility for Nature*, second edition, revised 1980, Duckworth, London.

Pepper, D. (1984) *The Roots of Modern Environmentalism*, Croom Helm, London.

Peters, R.H. (1977) 'The Unpredictable Problems of Tropho-dynamics', *Environmental Biology of Fishes 2*, 97–101.

Pettit, P. (1986) 'Social Holism and Moral Theory', *Proceedings of the Aristotelian Society 86*, 173–97.

Pimm, S.L., Rosenzweig, M.L. and Mitchell, W. (1985) 'Competition and Food Selection', *Ecology 66*, 798–807.

Popper, K.R. (1959) *The Logic of Scientific Discovery*, Hutchinson, London.

——— (1963) *Conjectures and Refutations*, Routledge and Kegal Paul, London.

Porritt, J. (1984) *Seeing Green*, Blackwell, Oxford.

Pough, F.H. and Andrews, R.M. (1985) 'Energy Costs of Subduing and Swallowing Prey for a Lizard', *Ecology 66*, 1523–33.

Price, P.W., Slobodchikoff, C.N. and Gaud, W.S. (1984) *A New Ecology*, Wiley, New York.

Prigogine, I. (1976) 'Order Through Fluctuation', in Jantsch and Waddington.

——— (1980) *From Being to Becoming*, Freeman, San Francisco.

Putman, R.J. and Wratten, S.D. (1984) *Principles of Ecology*, Croom Helm, London.

Quine, W.V. (1966) *The Ways of Paradox and Other Essays*, Random House, New York.

Rawls, J. (1972) *A Theory of Justice*, Harvard University Press, Cambridge.

Redclift, M. (1984) *Development and the Environmental Crisis*, Methuen, London.

Regan, T. (1981) 'The Nature and Possibility of an Environmental Ethic', *Environmental Ethics 3*, 19–34.

——— (1984) *Earthbound: New Introductory Essays in Environmental Ethics*, Random House, New York.

Richards, D.A.J. (1981) 'Rights and Autonomy', *Ethics 92*, 3–20.

Rolston III, H. (1975) 'Is There an Ecological Ethic?', *Ethics 85*, 93–109.

——— (1986) *Philosophy Gone Wild*, Prometheus, Buffalo, N.Y.

——— (1987) 'Duties to Ecosystems', in Callicott, J.B. (ed.), *Companion to a Sand Country Almanac*, University of Wisconsin Press, Madison.

Roszak, T. (1979) *Person/Planet*, Gollancz, London.

Bibliography

Routley, R. (1983) 'Roles and Limits of Paradigms', in Elliot and Gare.

Ruse, M. (1973) *The Philosophy of Biology*, Hutchinson, London.

——— (1986) *Taking Darwin Seriously*, Blackwell, Oxford.

Saarinen, E. (ed.) (1982) *Conceptual Issues in Ecology*, D. Reidel, Dordrecht.

Salmon, W.C. (1971) *Statistical Explanation and Statistical Relevance*, University of Pittsburgh Press, Pittsburgh.

Sartre, J.P. (1958) *Being and Nothingness*, (Barnes, H.E. trans.) Methuen, London.

Schumacher, F. (1974) *Small is Beautiful*, Abacus, London.

Schwartz, W. and A. (1987) *Breaking Through*, Green Books, Bideford.

Shoemaker, S. (1984) *Identity, Cause and Mind*, Cambridge University Press, Cambridge.

Simberloff, D. (1980) 'A Succession of Paradigms in Ecology', *Synthese 43*, 3–39 and in Saarinen (1982).

Singer, P. (1972) 'Famine, Affluence and Morality', *Philosophy and Public Affairs 1*.

——— (1981) *The Expanding Circle*, Oxford University Press, Oxford.

Skolimowski, H. (1981) *Eco-philosophy*, Marion Boyars, London.

Smart, J.J.C. and Williams, B. (1973) *Utilitarianism For and Against*, Cambridge University Press, Cambridge.

Sober, E. (1984a) 'Sets, Species and Evolution', *Philosophy of Science 51*, 334–41.

——— (1984b) *The Nature of Selection*, MIT Press, Cambridge, Mass.

Stevenson, L. (1974) *Seven Theories of Human Nature*, Clarendon Press, Oxford.

Stone, C.D. (1974) *Should Trees Have Standing?*, Kaufmann, Los Angeles.

——— (1985) '*Should Trees Have Standing?* Revisited', *Southern California Law Review 59*, 1–154.

——— (1987) *Earth and Other Ethics*, Harper and Row, New York.

Strong, Jnr, D.R. (1980) 'Null Hypotheses in Ecology', *Synthese 43*, 271–85, and in Saarinen (1982).

——— Simberloff, D., Abele, L.G. and Thistle, A.B. (1984) *Ecological Communities*, Princeton University Press, Princeton.

Suchting, W.A. (1983) *Marx: An Introduction*, Wheatsheaf, Brighton.

Sylvan, R. (1985) 'A Critique of Deep Ecology', *Radical Philosophy 40*, 2–12 and *41*, 10–22.

Tansley, A.G. (1935) *Introduction to Plant Ecology*, Allen and Unwin, London.

Taylor, P. (1986) *Respect for Nature*, Princeton University Press, Princeton.

Thomas, K. (1983) *Man and the Natural World*, Allen Lane, London.

Tilman, D. (1980) 'Resources: A Graphical-mechanistic Approach to Competition and Predation', *American Naturalist 116*, 362–93.

Tobias, M. (1983) *Deep Ecology*, Avant Books, San Diego.

Tudge, C. (1977) *The Famine Business*, Penguin, Harmondsworth.

WCED (1987) *Our Common Future*, World Commission on Environment and Development, Oxford University Press, Oxford.

Walker, L.R. and Chapin III, F.S. (1986) 'Physiological Controls over Seedling Growth in Primary Succession on an Alaskan Floodplain',

Ecology 67, 1508–23.

Warnock, G.J. (1971) *The Object of Morality*, Methuen, London.

Watson, A. and Lovelock, J.E. (1983) 'Biological Homeostasis of the Global Environment', *Tellus 3513*, 284–9.

Wenz, P. (unpublished) 'Species of Deep Environmentalism: A Critical Taxonomy'.

——— (1988) *Environmental Justice*, SUNY Press, Albany, NY.

Wheelwright, N.T. (1985) 'Fruit Size, Gape Width and the Diets of Fruit-eating Birds', *Ecology 66*, 808–18.

Whitham, T.G., Williams, A.G. and Robinson, A.M. (1984) 'Individual Plants as Temporal and Spatial Mosaics of Resistance to Rapidly Evolving Pests', in Price, Slobodchikoff and Gaud.

Wilbur H.M. (1984) 'Complex Life Cycles and Community Organization in Amphibians' in Price, Slobodchikoff and Gaud.

Williams, B. (1985) *Ethics and the Limits of Philosophy*, Fontana, London.

Williams, C.B. (1951) 'Intra-generic Competition as Illustrated by Moreau's Records of East African Birds', *Journal of Animal Ecology 20*, 246–53.

——— (1964) *Patterns in the Balance of Nature*, Academic Press, London.

Wilson, E.O. (1975) *Sociobiology*, Belknap Press, Cambridge, Mass.

Wilson, I.R. (1979) 'Explanatory and Inferential Conditionals', *Philosophical Studies 35*, 269–78.

Wimsatt, W.S. (1982) 'Reductionist Research Strategies and their Biases in the Units of Selection Controversy', in Saarinen.

Wollheim, R. (1984) *The Thread of Life*, Cambridge University Press, Cambridge.

Woodfield, A. (1976) *Teleology*, Cambridge University Press, Cambridge.

Wright, C. (1983) *Frege's Conception of Numbers as Objects*, Aberdeen University Press, Aberdeen.

Wright, L. (1976) *Teleological Explanations*, University of California Press, Berkeley.

Index

Index